DRINK THIS NOW!

JOHN BOYER

Cover art:

Front: Glass of Red Wine, Copyright ©2010 by Digital Vision
Back: Wineglasses, Copyright ©2010 by PhotoDisc, Inc./Getty Images

This custom textbook includes materials submitted by the author for publication by John Wiley & Sons, Inc. The material has not been edited by Wiley and the author is solely responsible for its content.

Copyright © 2010 by John Boyer.

To order books or for customer service, please call 1(800)-CALL-WILEY (225-5945).

Printed in the United States of America.

ISBN 978-0-470-61072-5

Printed and bound by EPAC.

10 9 8 7 6 5 4 3 2 1

Acknowledgements

Cover design and interior design: Katie Pritchard

Food, cheese, and wine pairing consultant: Trent Crabtree

Content Editors: Randall Horst, Joel Nachlas, Keith Roberts, Robert Tracy

Contents

intro

This is the introductory chapter to a book about wine. Since there are about a million or two books on this topic, I suppose I should follow the formula that all those other texts do. I should fill several pages of this intro glorifying the beverage itself, and describing its great role in history, and how important it is culturally, and maybe even giving you a rundown of how wine is made. I should wax poetic about how great wines are an art form, show you some statistics of wine production and consumption, and show you lots of colorful maps of wine regions of the world...

But I'm not going to.

See, this book is about drinking. Drinking wine to be exact. Drinking wine right now. Right this second. So before you even finish this intro, go to your local wine shop or grocer and buy a random bottle of wine. Don't know anything about wine and lack confidence to pick one out because it's all so confusing? That's cool. Just pick the bottle with the label that is the prettiest or most intriguing. It's a perfect way to start on this journey. Now bring that wine home, crack it open, pour a glass, and then resume reading.

Are you back yet? Good. Now drink some of that wine. Now I'll tell you why drinking that wine is so awesome. First, wine tastes good. Second, wine makes you feel good. Third, wine is good for you. Fourth, wine is good for business. Fifth, being a wine drinker makes you smarter. Sixth, wine helps you lead a richer, fuller life.

Those are bold assertions, aren't they? Ha! Well, that's what I'm good at...along with drinking of course! However, I'm not really trying to be cute or facetious in making these statements: I am a true believer that everyone should drink wine...and drink it now! Let's go through my short list again, to reinforce the themes of why you should be drinking this stuff now...

Wine tastes good Most of you non-wine-drinkers may be disagreeing with me already, as you may be thinking "Ewww..I don't like wine, it dries my mouth out, and taste so harsh!" But you will soon change your mind. You just don't know *how* to drink wine yet, which we will get to in the very first lesson. One of the fantastic things about wine is that there are so many different flavors, so many different styles and so many different variations from region to region and year to year, that every single person in the world can certainly find a wine that turns them on. You could drink a thousand different wines every single day for the rest of your life, and you still wouldn't get to them all! Saying you don't like the taste of 'wine' is like saying you don't like the taste of 'food'; how is that possible? And the more wines you try and the more knowledge you garner, the better and better wine gets…and yes, unlike any other beverage I know of, some great wines can border on orgasmic! Try getting that from soda pop!

Wine makes you feel good This beverage has alcohol in it, and as such has served as a relaxant and stress reliever for millennia. But unlike liquor or beer, we usually don't associate wine with over-indulgence. Wine is a social lubricant and, much like food, brings people together to share an experience. It is communal. It lowers inhibitions. It takes the edge off of uncomfortable situations. It helps conversation flow. We just feel better when drinking wine…in moderation of course. Which ties into…

Wine is good for you The stress-reducing elements of moderate wine consumption are not the only benefits to your health. It seems that study after study indicates that the tannins in wines (mostly reds) are extremely awesome for overall circulatory and cardiovascular well-being. Translated: wine is great for your circulation and for your heart. One need look no further than the folks who live in wine-producing regions, where lifelong integration of wine into daily diet equates to lower heart disease. But health is only one aspect of why wine is so good for you because….

Wine is good for your business And by business, I mean the business of living. I want to particularly pick on the younger generations who may be perusing this book. We are living in an era of transition in drinking habits here in the US, and I'll give you one guess as to how it's changing: that's right, wine consumption is going through the roof! We are finally morphing into a wine-drinking society, and that has repercussions mostly for you, my young friends. When you graduate from college and go for a luncheon job interview, your potential bosses are going to be drinking wine. When you move to the big city and snag a date with a classy girl/guy, they are going to be drinking wine. When you want to then impress that new boss or your new in-laws with a gift…it probably will be with wine. Are you ready for the challenge? Do you know your way around the wine world enough to impress for success? Or will you pass up opportunities by staying safe and ordering a Bud Lite? Drinking and being able to converse about wine puts you in a different league, the upper league, which brings us to…

Wine makes you smarter And what do you think of folks who drink wine anyway? When I ask this question in classes, I get the inevitable answers: people who drink wine are perceived as smarter, richer, better-educated, better-off, cultured, and sophisticated. Damn! Who wouldn't want to be identified with a group like that? Don't you? Understanding the dynamics of wine does seem to make you smarter…and not just about the product itself, but about cuisine and culture, and health and wealth. Being savvy about wine does make one—at least appear—very cosmopolitan.

Now I said wine makes you smarter, not snobbier. But have you ever wondered why we have this term 'wine snob'? Unfortunately, some folks in this upper-crust category use wine knowledge (or money, power, or knowledge of any kind) as a weapon to make themselves seem more important than the average Joe. Anyone who does that is a wine snob. They want to tout and spout their smarts about wine so as to make themselves look superior, and/or others feel inferior. And it has worked wonders over the centuries, as we do associate wine consumption exclusively with the high class. But, my friends, don't believe the hype! Wine is for us all! This book aims to help you demystify the beverage and disarm the snobs so you too can get in the game and enjoy all the benefits of this fantastic libation….which brings us to the last…

Wine helps you lead a richer, fuller life Add up the 5 reasons above, and you get this striking number six reason to drink wine now: life will be good. As a cosmopolitan, cultured wine connoisseur, you will be amazed at how much sweeter life can be. In my experience wine drinkers tend to be more relaxed, more interesting, more traveled, more engaging, and overall more healthier and happier. And I'm not talking about being part of an exclusive upper class here: I've partied with farmers and vineyard laborers and blue-collar workers the world over that are into wine, and they exhibit all the positive attributes as well. Wine is just such a democratic beverage…it truly benefits all who engage in its culture.
Wine drinkers pay more attention to flavors and foods, so that every meal becomes a joy onto itself. Imbibers of wine more often enjoy the simple pleasures of a sharing a glass with family or friends…or strangers. Wine knowledge is a lingua franca that is a conversation starter no matter where you are in the world. Wine is great. You should drink some now.

Wine is the beverage of moderation, of education, of socialization, and of civilization—and that is no exaggeration. Won't you drink this now?

What this book will do
Which gets us back to wrapping up this introduction. The title of this text is _Drink This Now_ which is just as much an encouragement as it is an instruction. This book aims to:
  ➢ Teach you how to drink wine
  ➢ Introduce you to wine drinking terms
  ➢ Introduce you to the different styles of wine
  ➢ Introduce you to the main grape varieties that make up most of today's wine world
  ➢ Provide a reference guide for food/wine experiences
  ➢ Create a vocabulary and understanding of wine, so you have a platform of basic knowledge onto which you can build for the rest of your life.

And we will achieve these goals by using actual wine drinking as a vehicle for the learning experience. See, the way you really learn about anything is to do it. Want to learn how to ride a bike? Get on it and peddle! The same holds true with wine…you got to get in the game and drink this stuff now in order to truly learn and understand it. It's why I wrote this text…each chapter and each lesson is essentially focusing a real wine drinking experience for you, with a bit of educational material on the side to encourage and enlighten you along the way.   If you are not prepared to drink this stuff now, then you may want to find another book. Maybe a nice book on origami would suit you better?

What this book will NOT do

However, this manual will not do a lot of other things that are important for furthering your wine education at the next level. This is an intro guide to drinking wine for the novice. As such, it focuses primarily on bringing you up to speed from knowing nothing about wine to being a confident wine drinker. A wine drinker that understands enough of the basic structure of the wine world so that you can smartly fill in details as you gain more experience. In essence, we will build the frame of your wine-drinking house in this book…and you can spend the rest of your life finishing that house and filling it with more stuff as you accumulate more wine experiences. Like what other stuff?

Other stuff you will eventually want to know, but this book won't do:

> For starters, this book *won't* teach you how to 'properly' assess a wine. Full-on wine assessment means looking and sniffing and sipping and spitting out the wine, and then writing everything down and assigning a point value system to judge it by. The hell with that. We want to drink the stuff.   Leave the 'proper' assessment to the wine writers, wine judges, the sommeliers, and the wine snobs. I'm a wine drinker. I only spit it out if it tastes like crap. Which is rare.

> This book *won't* deal with regional descriptions of wines. Granted, this is very important, but it serves to greatly confuse novice wine drinkers. Learning all the regions, and sub-regions, and sub-sub-regions that make wine, as well as what each of those wines is supposed to taste like, is a daunting task. But I don't want to undervalue the geography of a wine: location is of paramount importance.
>
> All wines are from somewhere, and that somewhere more often than not dictates the style and taste of the wine. Most wine books are structured by geography…they tell you about the wines of Burgundy, then the wines of Tuscany, etc. But I'm taking a different tack: we are going to focus on the grape varieties that go into making those wines as foundation block for your wine knowledge. Because, to be honest, the world is gravitating away from labeling wines by their region and moving towards labeling by their grape variety. Of course, I'll give hints about regional styles and geography along the way.

> It *won't* go into wine label interpretation. All those names and places and dates…and half of them are in a foreign language! And there are no set rules of labeling for the world, so everybody does it a little different. Damn, it's so confusing! Of course we'll be giving you hints along the way on this too, but this is another

book's worth of information in and of itself.
➤ It *won't* be a guide to wine price analysis or vintage evaluation.

All of those things are what makes the world of wine so damn complicated! And it is the appearance of complication that scares so many people away from trying wine...which is preposterous! You don't need to know the history and labeling conventions of the 1855 Bordeaux classification in order to appreciate how awesome a Bordeaux wine is. You can just drink it now! And that's what we are going to do in this book.

For those other really important things, and a whole lot more, I suggest using this drinking guide in conjunction with Kevin Zraly's *Windows on the World Complete Wine Course:* 25th Anniversay (ISBN 978-1402767678) and Hugh Johnson's *The Concise World Atlas of Wine* (ISBN 978-1845335007). Those are the complimentary texts that I use for my classes, and they will provide you a wealth of information to use for the rest of your wine drinking life.

By the time you finish this book, you will easily have the skills to hang in any wine conversation at any table; understand enough to confidently shop for wine in any store; and be able to impress your family and friends...and possibly even woo the date of your dreams.

Drink This Now doesn't necessarily strive to make you a wine connoisseur, a wine collector, a wine snob or a wino. But when all is said and done, it will make you a wine drinker. So if you are up for the wine challenge, turn the page to Lesson 1 and prepare to **drink this now.**

1.

PRELUDE TO DRINKING WINE

Don't be put off by the title! We are going to start our wine education right now, by starting drinking right now! However, this section consisting of three lessons which are a bit of background for why we drink wine, how we drink wine, and when we drink wine. Food, glassware, and aging/decanting all contribute to our wine experience, and as such should be addressed immediately before we get any deeper into our wine drinking exercises. Let's get to work!

Lesson 1. Why we drink wine with meals & how food alters the taste perception of wine.

Lesson 2. Glassware: Yes, it does make a difference

Lesson 3. When is it Time to Drink the Wine? Aging & Decanting and their affect on wine

Lesson 1: Why we drink wine with meals & how food alters the taste perception of wine

a.k.a. Why Europeans drink wine with meals, while Americans just drink straight up—it's the difference in styles man! Wine is food dude! Eat, drink, and see how flavors are married together in this lesson on the relationship between wine and food.

What to grab:
- One of the following from your local wine shop or grocer. If they don't have these exact bottles...no worries! Just ask for some assistance picking out a wine similar to one on this list:
  - **Bottles $5-$15**
    - ➤ Sant Agata Altes Barbera (Italy)
    - ➤ Domaine Grand Veneur Cote du Rhone (France)
    - ➤ Fossi Chianti Classico (Italy)
    - ➤ Castell del Remei Gotim Bru (Spain)
    - ➤ Erial Epifanio Ribera Del Duero (Spain)
  - **Bottles $16-$30**
    - ➤ Jaume Clos de Sixte Lirac (France)
    - ➤ Toscolo 04 Chianti Classico Reserva (Italy)
  - **Bottles $30+**
    - ➤ Fontodi Chianti Classico (Italy)
    - ➤ Erial TF Ribera del Duero Reserve (Spain)
- A hard Italian cheese, like Parmesan or Asiago
- A semi-spicy or full-on spicy food, like salami, pepperoni, bacon, pizza or a steak. Hell, just have it with your dinner.

What to do:
1. First open the bottle and try immediately. What is its color? What do you smell? What flavors do you taste? Write down some descriptions of your experience, both what you smell and how it tastes.
2. Next wait at least 2 hours for the wine to "open up," leaving the cork off the wine allows oxygen to get in. Try the wine again. Has the smell changed? How so? What about the taste? Is it fruitier? Is it more appealing to you?
3. Now try the wine with cheese. And I mean try it: don't be shy or proper. Get a big ol' hunk of that cheese and start chewing on it, and then start sipping the wine while you work it through. Record your reactions. Has anything changed in the taste? Does the wine seem smoother or lighter or fruitier? Repeat with a different type of cheese if available. Again, detect some differences and make yourself write down some different descriptors.
4. Now try the wine with food. Does the spiciness of the meat bring out any different flavors in the wine? Record your reactions again. If you are going for the gusto with a full fledged dinner, try a bite of each individual food with a sip of wine and note any differences

of taste on paper. Does the meat with wine create a different taste than the asparagus or potatoes with wine?

5.   Finally (if possible), save a bit of wine in the bottle. Let it sit exposed (no cork) on the counter for a few more hours. Let it go overnight if you want. Then try it again on its own, without food. What's it like now? Write it down. Compare this with your first impression of the wine right after you opened it. What has happened?

What to look for:

I'm not a big fan of putting words in people's mouths, but it is usually helpful to the 'uninitiated' to have some descriptive words to choose from as you first start this process. Here is a small arsenal to choose from as you go thru this exercise:

- **Just opened wine:** hot, spicy, mouth-puckering, acidic, tannic, tart, cedar, cocoa, rough, aggressive
- **With cheese:** fruity, jammy, plum, cherry, licorice, blackberry, raspberry, softer tannins, smooth, silky, etc.
- **With meat:** earthy, dry, spicy, cedar, smoky, flinty, peppery

Don't search for these descriptions. Just let it happen. Whatever terms apply best to what you are smelling and tasting is what you should use.

Why this is:

A couple of different things are going on here that make the smells and tastes change over time, and with different foods. Let's start with the food part.

Generally speaking, European wines (or 'Old World' wines, as a lot of us refer to them) are crafted much more to go with food than are their 'New World' (California, Australia, South America) counterparts. New World wines are typically fruity and silky right from the start, all the way to the end of the bottle. These are definitely traits to be admired, because they make the wine immediately accessible and pleasant to the palate—which is why people drink them all the time, anywhere, with or without food. And I'm not complaining either; I'll take a powerhouse Cab Sav or Shiraz from the New World any time it's offered, food is optional!

On the other hand, Old World wines typically are balanced between acidity, fruitiness, and earthiness. The acidic or sharp edge is usually

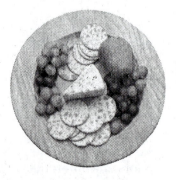

detected right when you start drinking it, especially without food. But a funny thing happens when you add any sort of fatty material in the form of cheese or meat: the acidity melts away, exposing a lot of the fruity flavors underneath. When other intense flavors are introduced, such as spiciness or pungent ingredients, the earthy undertones also come out to play. Indeed, most Old World wines cannot fully express themselves without the pairing of foods. And that's the beauty of it, because different foods interplay in different ways with the wine to create different flavors. As you may know, European cuisine is very local as are their wines, meaning that every little town and region has specialty cuisines that have evolved over centuries—as have the wines that go with them. Thus, the pairing becomes essential to fully 'get' the wines from these areas.

The other thing going on here is time. A very young wine, and/or a wine just opened, can be 'closed' to the senses in that it is not expressing all of its flavors just yet. When this happens, the tannic, dry, tart, acidic nature seems to jump out at you. After a wine is open and exposed to oxygen, it is in essence starting to age at an accelerated pace. Leave it open for too long, and the oxygenation leads to spoilage. But as your bottle sits open for a couple of hours or overnight, it is in essence aging to perfection— albeit very rapidly. This airing out, or 'breathing' as you've probably heard reference to, essentially softens those tannins and acidic edges, which produces a noticeable difference on the palate and perhaps even changes the descriptions you give to the smells.

Anytime you can, test out this theory by leaving a bit in the bottle after your meal and trying it the next day…or two days later, or whenever. See how far you can push the wine. Be forewarned: most white wines will be totally shot within 24 hours without refrigeration, and most reds will not make it much further than that on the counter. It does seem to me that better crafted wines (usually meaning more expensive ones) can easily make it overnight and beyond, and be even better for the wear.

If the wine did not seem to change at all to your palate, no matter what food you sampled or how long you let the wine sit: don't be discouraged. This stuff takes time. Be patient, and try again with another wine. Pay attention to what you smell, and what you taste. Try your best to form descriptions, and then keep trying, and then keep trying after that. Attaching words to smells and flavors is not an easy task with wines, because so much is going on and it keeps changing on you. Orange soda just tastes like orange soda. French fries just taste potatoey. Chocolate just tastes chocolaty. A wine can have ten smells and tastes, and then ten different smells and tastes after a piece of cheese….which is hopefully what you just found out.

Name of wine:	
Year:	
Region:	
Color:	
	Comments
Just opened	
Bouquet (smell):	
Taste:	
Body (heavy, light):	
Other comments:	
Two hours later	
Bouquet (smell):	
Taste:	
Body (heavy, light):	
Other comments:	
With cheese	
Taste:	
Body (heavy, light):	
Other comments:	

	Comments
With meat/meal	
Taste:	
Body (heavy, light):	
Other comments:	
A day later	
Bouquet (smell):	
Taste:	
Body (heavy, light):	
Other comments:	

Lesson 2: Glassware: Yes, it does make a difference

How the shape of the glass affects your perception of the wine. Some typical glass shapes for some typical types of wine. Why some people hold their glass by the stem while others palm the whole bowl. Why people do all that swirling nonsense. What are the processes involved.

What to grab:
- As many different shaped glasses as possible, but at least these:
  - ➢ a shot glass
  - ➢ a short, stout, typical 'rocks' glass
  - ➢ a tall, thin, typical water glass
  - ➢ a flared, smaller wine glass
  - ➢ a flared, big-as-you-can-get, wine glass
  - ➢ a fishbowl (seriously, do it if you want, and send me a picture)
- One bottle of each of the following from your local wine shop or grocer:
  - ➢ Any aromatic white wine like Gewürztraminer, Viognier, or Albariño
  - ➢ Any aromatic red wine like Shiraz or Zinfandel
- Bread or unsalted crackers. Plain, with no fancy flavors attached.

What to do:
1. Throw the bottle of white wine into the fridge for 30 to 45 minutes, until it has a good chill on it. While you wait, get out your glasses, and go ahead and pull the cork on the bottle of red. Line up your glasses as listed above. Got more shapes and sizes? Stick them in the line-up at your discretion but roughly going smallest to biggest.
2. Pull out the white wine and use the shot glass as a measure. Put one shot of the white wine in each of your glasses. Fill the shot glass when you are done, and put it at the head of your line. Now you are ready to go.
3. Time for a sniff-a-rama. Pick up each container, one at a time of course, and give it a good sniff. Start with the shot glass itself. Get it as close to your nose as you dare, take just one good deep sniff, and then put the glass back on the table. Record any and all of your impressions of the nose on that wine. Take at least a minute break, and then repeat the same process for all your other glasses.
4. After recording all impressions of the nose of that white wine in

each glass, its time to drink it.   But not in the same order! This time, sniff and then drink the wine in the shot glass first, and record your impressions. Next, go straight to the flared smaller wine glass, sniff and then drink it down. Now go to the biggest flared wineglass/ fishbowl and give it a whirl. As always, record impressions of the nose. Are you able to detect some differences yet? Is the same wine more/less pungent in different containers? Can you taste/smell new flavors according to the vessel?

5.   Time to hit up the red wine. Drink the rest of your white wine, or just chuck it out, and set up same scenario as Step #2. Line up the glasses and put a shot of your red wine in each. Repeat the tasting exercise steps #3-4. Take good notes on your tasting sheet along the way.

6.   Before we finish, let's do one more thing. Take the red wine left over in your other glasses/bottle, and pour some of it into your small, flared wine glass and the rest into your large, flared wine glass, or your fishbowl if you have one. Leave the small glass on the table, but pick up this big flared wine glass. Cup the bowl of the glass in both hands, bringing as much surface area of your hands in contact with the glass as possible…and preferably right where the wine is inside the glass. Gently slosh the liquid around the inside of the glass, making every effort to 'coat' or 'paint' the inside of the glass with wine. Do this for several minutes. Now set that big glass down and pick up the smaller glass. Do your  smell assessment on it, wait a minute, and then immediately pick up the big glass and repeat the smell assessment on it as well. Has anything changed? Does the exact same wine in these two different glasses smell any different? Now go ahead and do the taste assessment between the two glasses. Any difference there? *How can the same wine in different vessels present so much more smell…and possibly even more taste?*

What to look for:

Hopefully you discovered that not all drinking vessels are created equal… at least not when it comes to getting the most out of your wine. You probably found that the smaller the glass, the less of the wine aromas you could detect. But it also has something to do with the shape of the glass too. Short glasses and tall, thin straight glasses don't seem to focus the wine aromas as well as those with rounded and tapered shapes.

And what happened when you 'warmed up' that red wine in the big glass by holding and swirling it? The aromas should have been even more pronounced than ever. Intensifying the aromas may have even transferred over to the taste department as you drank it. Did it?

Why this is:

When we taste wine, or really when we taste everything, so much of what we perceive to detect on our palate is actually going on in our nose. Huh? That's right: most of what we 'taste' is actually what we smell. Some say almost half, some say more than half…hell, I say almost all…taste perception occurs in the schnozzle. So much of the aromas and flavors of wine are actually recognized and appreciated by the nose, that without our sense of smell, wine would be no different than water…well, except for the sweet buzz that you certainly can detect with no help from your nose at all.

Doubt me? Well, do you remember the last time you had a head cold and your nose was all stopped up…how did food taste to you? It didn't taste like anything. You don't have to wait for sickness to strike to test this theory: just pinch your nose shut right now and eat some pungent cheese. You won't be able to detect a thing except the mouth feel or texture of the food itself.

But wait; what the hell does a head cold have to do with the shape of a damn wine glass? Just this: it's all about the sniff, baby! You got to smell it to enjoy it, and some glasses are simply better at helping out our sense of smell than others are. So what shapes are best?

Why shape matters

Before I go any further, let me make one point expressly clear: You can drink any damn wine you want out of any damn glass you want. My main job is to demystify wine, and I never want to be accused of snobbish habits, or of telling you that there is only one single 'right' way of enjoying wine. I have consumed wine out of every vessel imaginable, including some articles of women's apparel, and always could appreciate the wine to some degree. The only truly bad glass for wine is one that leaks!

If you ever party in the 'Old Country' of Europe, you will often see old men in small Italian villages drinking wine from old jelly jars, or you may even be served a fine sherry in Spain in a shot glass. You can go to a supermarket in France and buy wine in a plastic gallon jug (just the same way we buy milk in this country) for just a buck or two, and serving that in a water glass would be just dandy. It might not help that wine much to be in a better glass.

However, some glass shapes do greatly accentuate the aromas and flavors of wine… so if you are having a good wine, you probably want to get the most out of your experience. Right? So you should use a glass that is best

suited for the purpose. And what is that purpose? Answer: Concentrating the wine molecules.

When we smell anything—be it someone's perfume or freshly cut grass or even food we are chewing in our mouth—we are sucking in molecules of moisture into our nose which transmit information about the item, which in turn is transmitted to the brain and stored in long term memory. If the code matches up with something we've smelled before, then we recognize it and can describe it! It is important to note that these molecules move around via moisture...which is why you can smell a skunk 20 miles away on a rainy night. And why cut grass is so pungent just as it's freshly mowed. And why the glass shape matters...

See, the best shaped glass for wine is definitely the tapered bowl that you used for this exercise. This shape does a couple of things simultaneously:

1) The bowl shape increases the surface area of the glass that the wine can come into contact with. Increased surface area allows more of the wine moisture molecules to evaporate and enter the air just above the liquid as well as above the glass surface which has been coated with wine as it sloshes around. More wine molecules in the air = more molecules we can sniff in and assess.

2) The narrow mouth of this glass focuses those moisture molecules into a narrow area which you then stick your nose into and sniff up. Consider that tapered neck as an upside-down funnel which is siphoning those wine molecules straight into your nose.

There are always exceptions and variations of course. Champagne flutes are long and narrow because their primary goal is not the concentration of aroma, but the presentation of bubbles. No, I'm not kidding. Bubbles are what make that wine so special, so you want to be able to see them floating up in their long lines to the top of the glass. Sherry glasses are typically small narrow little cordial glass affairs based mostly on the typical small serving size. And brandy and port glasses are typically those squat fishbowl shapes which allow for long, luxurious sniffing and swishing traditions.

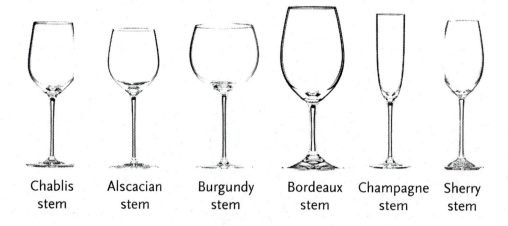

Chablis   Alscacian   Burgundy   Bordeaux   Champagne   Sherry
stem   stem   stem   stem   stem   stem

So why don't the other shapes in our exercise work as well? Standard water glasses, mugs, or rocks glasses lack that focusing factor which the tapered glass brings. In those glasses, the wine molecules may enter the air, but are quickly dissipated straight up and out, much like smoke out of a chimney.

And the worst case scenario was that shot glass: it had no containment of aroma molecules above it at all! Molecules up and out, non-stop. Plus, since it was so completely filled with liquid, the surface area available for evaporation of the wine was minimal…limited only to the surface of the wine itself, which of course was really small. And small is a good launching pad for…

Why size matters

As just elaborated upon, the small shot glass seriously limited the surface area of the wine exposed to allow the wine molecules to get into the air for us to smell. And you can extend that thinking all the way up the line for any shaped glass: the smaller it is = less surface area = less wine molecules get into the air = less smell. Even though the shape was not optimal, you probably could detect a lot more of the wine's aromas in the bigger water glass than you could in the rocks glass, or the shot glass.

Even when dealing with the primo flared wine glass, size helps here too. A smaller, more narrowly tapered wine glass (usually reserved for whites) won't bang out as much aroma as some of those monster bowl wine glasses (usually reserved for some very special red wines) you may have seen at a fancy dinner. But no one wants to carry around a gigantus fish bowl every time they have a glass of wine, so most stemware trades off maximum size for versatility and utility.

However, I must admit to you, those big-ass glasses do bring out the best in serious wines…and can even help mediocre wines seem fuller and richer. There is something to the snobbery here…just as there is to what is perhaps the most pompous-looking performance in wine swilling, which is of course that wine swirling!

Why swirling matters

I know, I know, I know…I mock the wanna-be wine snobs all the time, and nothing seems to cry out 'snobby' as some highbrow at a cocktail party swirling his wine around in the air. And I know that as novice drinkers, you may be put off by the practice, or intimidated by watching people swirl and sniff. But people, let me be the first to tell you: you gotta do it! Especially for those reds! Perhaps an explanation of why we do it will help. It goes back to that surface area stuff I just talked about above. When you want to get the most out of that wine, you need to put as many of those aromas and flavor molecules into the air in the glass, so that you can sniff them in. Having a big glass helps, and having a good shaped glass helps more, but you can assist in the molecule movement too! Swirling the wine coats the interior of the glass, which in turn increases surface area of the wine, which in turn puts more of those molecules into the air.

Need a negative example of this principle? Consider a small bucket with paint in it. You can probably smell the paint if you are right beside it, but probably not if you are on the other side of the room. Now what happens when you spread that paint over every wall of that room? Damn! It stinks! And bad! Those paint smell molecules are all over the place. But wait… the amount of paint is exactly the same, whether in the bucket or on the walls. Only the exposed surface area has changed, and changed radically. So you want to 'paint' the inside of your glass with wine, to get the same effect.

Finally, there is one last thing we can do to further increase the wine molecule count in the glass…and that is to heat it up! Just as boiling water causes evaporation which puts water molecules in the air above the pot, we can use our hands as 'heaters' to raise the temperature of the glass. That is why you will see folks cupping the bottom of their red wine glass, in effect 'boiling' the wine and putting more molecules in the air. The palming of the wine glass bowl often goes hand-in-hand with the swirling, thus using both practices to maximize the molecule count.

On the opposite front, we typically like to keep our white wine chilled so most folks hold their white wine glass by the stem, or even by the base. Its also why 'standard' white wine glasses are a bit smaller than reds…because we want a smaller pour of the white wine, thus leaving more wine in the bottle on ice or in the fridge in order to maintain the chill. Are you chill with that?

In summary:

- ❖ Most taste is actually perceived through smell, thus most of our enjoyment of wine comes from smell. This is why we spend so much time sniffing it, and why some glassware helps accentuate this process.
- ❖ The tapered shape of a typical wine glass helps increase the surface area of the wine and concentrates the wine molecules for you to sniff out.
- ❖ Bigger wine glasses do increase the surface area more, which increases the wine molecules in the air more too.
- ❖ Swirling the wine and 'warming' the wine also increase wine molecules for us to smell. Since we typically consume white wines chilled, the 'warming' practice is largely done just to red wines.
- ❖ Different wine glass shapes have evolved for lots of different types of wines, but they all pretty much serve the same basic functions.

Glass shape, glass size, swirling, warming…all for the sake of those wonderful molecules that we can smell! But damn, isn't it worth it?

Glassware

	Shot Glass	Rocks Glass	Water Glass	Small Wine Glass	Big Wine Glass	Fishbowl
White						
Red						

Wine

Lesson 3: When is it Time to Drink My Wine? Aging & Decanting and their affect on wine

Why we age some wines, and not others. Why we decant some wines before serving. What are the processes involved. What wines should we drink now. What wines we should age…and for how long.

What to grab:

This isn't really a cheapie experiment. Feel free to work with others, but it is essential for the lesson to get three of the exact same bottle of wine. If you can't find the exact wines listed below, just ask the folks at the wine store to hook you up with any young 'Old World' tannic red that should age out 5 to 10 years. They will know what you're talking about.

- Grab three bottles of just one of the following from your local wine shop or grocer:

   Bottles $5-$15
    - ➢ Sant Agata Altes Barbera (Italy)
    - ➢ Domaine Grand Veneur Cote du Rhone (France)
    - ➢ Fossi Chianti Classico (Italy)
    - ➢ Castell del Remei Gotim Bru (Spain)
    - ➢ Erial Epifanio Ribera Del Duero (Spain)

   Bottle $16-$30
    - ➢ Jaume Clos de Sixte Lirac (France)
    - ➢ Toscolo 04 Chianti Classico Reserva (Italy)

   Bottles $30+
    - ➢ Fontodi Chianti Classico (Italy)
    - ➢ Erial TF Ribera del Duero Reserve (Spain)
    - ➢ Guigal Chateauneuf Du Pape Rouge (France)

- A decanter, if you have the means, or any large container like a pitcher
- A small funnel
- A hard Italian or Spanish cheese, like Parmesan or Asiago or Peccerino

Don't worry if you can't find these exact labels. You can always ask your local wine store salesperson to help you pick out a good representation of the style I have described above. Show them the list and tell them to get you something close to it.

What to do:

1. With a pen or sharpie, mark one of your bottles "24 HOURS", mark the next one "12 HOURS" and mark the third bottle as "NOW."
2. This is a long-term lesson which actually will start the day before you actually do it, so plan ahead. A day before you want to drink, open up the bottle labeled "24 HOURS" and decant it. Decanting for this lesson simply means putting the wine into a bigger container, like a pitcher. Pour the whole damn bottle in, slosh it around a little bit so that it gets some air mixed into it, and then

use your funnel to pour it back into the bottle. That's right, pour it back into the bottle but do not replace the cork. Set it on the counter. Despite the title of this book, resist the urge to drink this now.

3.   Roughly 12 hours later, repeat step #2 on the bottle labeled "12 HOURS." Again, resist the urge to drink this now.

4.   Roughly 12 hours after that, pull out your glasses and the two opened bottles, and now pull the cork on your third bottle labeled "NOW." Do not decant this one. It is finally time to Drink This Now.

5.   In three different glasses, pour a healthy sample from each of the three bottles. Keep the glass right in front of the labeled bottles so you can keep them straight.

6.   Pull out your assessment sheet and go to work. Start with NOW, then move to 12 HOURS, and finish with 24 HOURS. Take a big sniff or two. What do you smell? Then take a big sip and slosh it around your mouth for a few seconds before you swallow. What flavors do you taste? Write down as many descriptions of your experience as you can muster, both what you smell and how it tastes. Take a minute or two break between each wine to rest your senses.

7.   Take a ten minute break, and then repeat that last step, but just with the smelling part. Go back and forth from the NOW to the 24 HOURS directly at least once. Detect any differences?

8.   Now try the wines with cheese. As in Lesson One, don't be shy. Get a piece of that cheese and start chewing on it, and then start sipping the wine while you work it through. Record your reactions for each wine. Has anything changed in the taste? Do any of the wines seem smoother or lighter or fruitier with the introduction of food? Which ones benefit the most by the addition of food?

What to look for:

As wines grow older, they lose a bit of their 'edge' or 'bite' that is provided mostly by acids and tannins. Now with most white wines, and even a lot of bulk red wines, this is a fairly negative process, as losing their edge means they are losing the good qualities that make them desirable. However, with young, tannic, aggressive red wines which are crafted to last, this loss of 'bite' over time makes the wine more drinkable.

Over time then, red wines 'soften' and become more approachable. But we don't necessarily have to wait a decade to achieve some of these results. What you have seen in this little experiment is that opening and decanting the wine just for a day accelerates this aging process. You should have been able to detect a significant 'softening' of the wine that had been open 12 hours, and even more so the one open for a whole day, when compared to the newly opened bottle. Well, did you? Why is this?

Why this is:

As referred to in Lesson One, a couple of different things are going on here that make the smells and tastes change over time, and with different food. Let's start with the food part.

Generally speaking, many European wines (or 'Old World style' wines, as a lot of us refer to them) are crafted much more to go with food than are their 'New World style' (California, Australia) counterparts. New World wines are typically fruity and silky right from the start, all the way to the end of the bottle. These are definitely traits to be admired, because they make the wine immediately accessible and pleasant to the palate—which is why people drink them all the time, anywhere, with or without food.

On the other hand, Old World style wines typically are balanced between acidity, fruitiness, and earthiness. On top of that, many wines are crafted with a high tannic component right from start. A tannic component? What the hell is that? Tannins are the stuff in grape skins, grape seeds, grape stalks, and even in oak barrels that impart a lot of body and distinctive taste to the wine. And that taste is astringency: a bitter, sometimes hot and harsh, mouth drying characteristic that makes you pucker when present in high amounts.

The tannins and acids in young red wines makes for a sharp, bitter or aggressive edge—which is usually detected right when you start drinking it, especially without food. But a funny thing happens when you add any sort of fatty material in the form of cheese or meat: the tannins and acidity melts away a bit, exposing a lot of the fruity flavors underneath. Heavy tannins are also in really strong black or green teas, if you've ever had those. And what do a lot of people add to tea to make it more drinkable? Cream, or fatty milk! Same process my friends.

So adding food softens this tannic balance. But the other thing going on here is time. Slow chemical reactions which occur in the wine over time to be more scientifically precise. Over time, as a wine ages, tannins and acids soften naturally resulting in a more approachable beverage. The harshness, bitterness, and edginess dissipates as the tannic component fades. A wine that is aggressive in its youth will mature to something much rounder and softer as it ages.

A very young wine, and/or a wine just opened, can be 'closed' to the senses in that it is not expressing all of its flavors just yet. When this happens, the tannic, dry, tart, acidic nature seems to jump out at you. After a wine is open and exposed to oxygen, it is in essence starting to age at an accelerated pace. Leave it open for too long, and the oxygenation leads to spoilage. But as your bottle sits open for a couple of hours or overnight, it is in essence aging to perfection—albeit very rapidly. This airing out, or 'breathing' as you've probably heard reference to, essentially softens those tannins and acidic edges, which produces a noticeable difference in the aromas and perhaps even changes the sensations on your palate as you drink the wine. This intentional 'opening up' or 'breathing' of the wine is what we refer to as decanting.

Why we decant

This section can now be summarized fairly succinctly now. We decant wine primarily to accelerate the aging process right before our eyes. By pouring wine into a bigger container like a decanter (but really anything will really do) we are introducing oxygen into the liquid, in effect making it come alive and soften up and become accessible.

Some folks will also decant in order to filter out solid particles. As big tannic red wines age they are chemically changing. In this process, some particles fall out of suspension (precipitate) and collect at the bottom of the bottle. While these particles are completely natural and completely harmless, most folks like to siphon them off in order that their guest won't accidently get a mouthful of what appears to be dirt. And if you get a big enough mouthful of the stuff, it kind of taste like dirt too. Ew.

Honestly though, there are few wines which benefit greatly from decanting. And that's because there is not a big percentage of wines which are age worthy. Did you say age?

Why we age wine

So if we can just decant the wine, and thus age it and open it up immediately...then why the hell would we ever age any wines? I mean, think about it. If you were a winemaker, would you want to create a product that has to sit in a barrel for a couple of years, and then sit in your warehouse and age for another 8 years? Which means you have to wait for ten years before you start to sell it and have a return on your investment. Can you think of any other business that wants to sit on its products for a decade? It's madness!

In fact, why the hell would a winemaker even put so much tannin and acid in the wine in the first place? Why can't they just make the wine ready to drink right this second? Well....most do nowadays. What we refer to as 'New World style' indicates that the wine has a level of smoothness and approachability as soon as it is bottled, no extra time needed. This is the trend with wine producers across the globe right now. It does make more business sense, as you can move the product immediately. The consumer doesn't have to wait for the wine to integrate or smooth out. They can drink it right now.

However, there are those wines that are still crafted intentionally to be aged for a while, and sometimes a long, long while. See, those wine makers don't extract tons of tannins and put them in the wine just to piss off your palate. They do it to make the wine last...but why?

Quite frankly, it's still a bit of a conundrum. See, some wine flavors and wine harmony can only be achieved by extremely slow chemical reactions occurring over a long period of time. Slow aging brings forth flavors and textures which simply cannot be duplicated in any other way. I know it sounds mystical and mysterious, but that's the way it is. That's what makes those great wines so great! And think about it: if a winemaker could (right this second) mimic the exact tastes and textures of a 50 year old Burgundy that sells for $5000…wouldn't he do it? Somebody would certainly try to do it…but they really can't. Some things only happen with time.

Only time can soften tannins and acids, integrate the wine's components, and meld and form flavors which together achieve awesomeness. Consider it a 'slow burn' of integration and evolution of the wine.

Winemakers going for this brass ring of awesomeness intentionally build up high amounts of tannins and acids in order to give the wine longevity enough to go the distance to achieve this goal. But be forewarned…not many winemakers in today's wine world are going for this effect. And their numbers get smaller every year…

So what wines do we age…or decant…and for how long?
Almost all folks who are unfamiliar with drinking wine maintain this basic assumption: the older a wine is, the better. WRONG! Nothing could be further from the truth. I suppose this bit of urban legend evolved from folks looking at the prices of the most expensive wines in the world… things like aged Burgundies, aged Bordeaux, and aged Barolos.

Yes, those wines are really expensive. And yes, part of the reason they are so expensive is because they are so old, and so rare. So I guess people put 2 and 2 together and end up with 27 ¾. Because it just ain't so that aged = better. The reason that those particular wines are allowed to become so old is because they are **age worthy**…meaning that they are crafted specifically to evolve over time. They actually need the time in order to become the masters of the universe of wine that they are.

But make no bones about it: most wine produced today is meant to be consumed young. Let's break it down:

Whites
Virtually all white wines are built for immediate consumption; the younger, the better. Whites are mostly built to express the fruity characteristics of the grapes they are composed from, and often with a slight acidic backbone to give them some 'umph.' Consider them the closest thing to the fresh fruit itself. And how long will fresh fruit last on the counter? That's right…eat it quick! Same with the white wine. Its best fruit aromas and flavors are expressed soon after creation and bottling. Time is no friend to the light, crisp, fruit flavors of white wine. Most white wines will be losing their luster fast after a couple of years. Unless the winemaker has done something to beef it up…

How to add some additional flavors and longevity to whites? Barrel age it to add some tannins. Or craft it with lots of sugars or lots of acids, or both. As you will see in a later lesson, beefing up the acids or the sugars to high levels makes even white wines more resistant to deterioration or spoilage due to bacteria. So white wines that can hold their own more than a few years include very acidic whites like Chablis or New Zealand Sauvignon Blancs, and very sweet whites like some German Rieslings which can last for decades and beyond. But to restate, those long-lasting whites are a rare beast…like 99% of whites produced today are ready to drink right now.

Reds

Believe it or not, a lot of red produced nowadays are best consumed fairly soon after being bottled too. Most are good to go as soon as they hit the stores, many will get better for just a few years, a smaller percentage than that will be best in 8 to 10 years, and only a handful of red wines are really crafted specifically for the long haul. However, unlike their pale brothers, all reds have a tannic component which is why even wimpy red wines last much longer than whites do.

But which reds benefit from aging or early decanting then? This is where the fine art of building a wine cellar and buying wine specifically for aging comes into play, but I'm not going to get into all that complicated stuff in this book. Just know this for now: 'Old World style' crafted red wines made from specific grapes are usually the ones that go long for the big score. These wines when young are typically deep red to inky black in color, have a high tannic component, high acidic component, and can be abrasively harsh and mouth-drying. A few examples of 'big' grapes that add that kind of firepower include Cabernet Sauvignon, Nebbiolo, Zinfandel and Tempranillo.

As you experienced in this lesson, when you come across a red wine that seems over the top-abrasive-harsh-hot-astringent, you can always decant and let it open up and mellow out for a while. You should now also be figuring out this now evident bit: that wine is also a good candidate for extended aging.

Big Alcohol Bombs

And I should just mention the last big category of wines, because they have the most aging potential of them all. All wines which have very high levels of alcohol can last for extended periods of time. These wines obtain high alcohol in a variety of ways which we will discuss in a later lesson, but know for now that Port, Sherry, and Madeira can easily last for a decade or two…or ten!

The high alcohol essentially prevents any bacterial spoilage indefinitely, and some of these wines are intentionally oxidized, which means that the introduction of air won't make them go bad either. So you can open these wines and pour a glass, and then leave the rest on your counter until you are ready for another glass in a week, a month, or a year. In addition, these wines have high levels of sugar or tannins (or both!), which also need time to integrate with the high alcohols in order to mellow out, so they won't even be 'released' for sale for years after their initial production. A high quality vintage Port is typically aged for 20 years before its ready to be opened. Talk about 'opening no wine before its time'!

Summary
- ❖ Most wines are produced for immediate consumption.
- ❖ However, some 'Old World style' reds are crafted with high tannic and acid structure which makes them harsh and abrasive when young, but gives them age-worthiness.
- ❖ The tannins and acids in such wines can be softened in two different ways: decanting or aging.
- ❖ Age worthy wines typically have high sugars or acids for white wines; high tannin and acids in red wines. High alcohol levels also increase longevity.
- ❖ Some special wines pick up flavors and harmony with age that can only be achieved with the passage of time.

Yes, there are those special wines that take their own sweet time to become magnificent works of drinkable art! Luckily, we have boat-loads of immediately quaffable wines produced every day to keep us busy while we wait for the great ones to bloom! It's like sitting in a big field of daisies and carnations waiting for the rare orchid to bloom…but with a buzz! Isn't wine awesome?

Name of wine:	
Year:	
Region:	
Color:	
	Comments
Just opened	
Bouquet (smell):	
Taste:	
Body (heavy, light):	
With Cheese:	
12 hours later	
Bouquet (smell):	
Taste:	
Body (heavy, light):	
With Cheese:	
24 hours later	
Bouquet (smell):	
Taste:	
Body (heavy, light):	
With Cheese:	

2.

WINEMAKER
INFLUENCE

Now comes the real work...as if you can call any activity that involves drinking wine as 'work.' But we are going to learn a lot in this section about how wine is crafted, with an eye towards how the art of the winemaker translates to the finished product. I am not terribly concerned with individual winery practices, and I flat out don't care about the intricate details of wine-making at a biological level. However, I am very keen to having you comprehend the major components and keywords of wine that you will hear in wine discussions, see on wine bottles, and read about in wine descriptions.

Oak, sweetness, acids, color, tannins, alcohol, body, balance, and mouth feel are elements of every wine that we need to know about, recognize, and understand. Why? So that we can appreciate wine when it's good, differentiate factors in wines that are not, and mostly to be able to intelligently discuss wine with others. While the lessons in this book are stand-alone and can be done in any order you wish, I would highly advise doing these four lessons first before proceeding any further. Let's get to it and do it...

Lesson 4. Got wood? Recognizing the role of oak in wine

Lesson 5. Dry vs. Sweet: The Sugar Showdown

Lesson 6. Color Me Clueless: White vs. Rose vs. Red Wines

Lesson 7. Rock Your Body, but Keep Your Balance! Light, Medium, and Full-Bodies Wines, and What That Means

Lesson 4: Got Wood? The influence of oak on wine

The affects of the oak barrel on wine. What it tastes like. How it affects the wine. What are the processes involved.

What to grab:
- One bottle of from each of the following lists:
  - ➤ Jewel 'Un-oaked' Chardonnay (California) or
  - ➤ Yalumba 'Unwooded' Chardonnay (Australia) or
  - ➤ Tolosa Chardonnay No Oak (California)
      AND
  - ➤ Toasted Head Chardonnay (California) or
  - ➤ Butterfield Station Chardonnay (California) or
  - ➤ Trapiche Oak Cask Chardonnay (Argentina)
- Bread or unsalted crackers. Plain, with no fancy flavors attached.
- Water

Don't worry if you can't find these exact labels. You can always ask your local wine store salesperson to help you pick out a good representation of the styles I have described above. Show them the list and tell them to get you something close to it.

What to do:
1. First, before you even chill it, open both bottles immediately and try a glass of each. Describe the color of each. What do you smell? What flavors do you taste? Write down some descriptions of your experience, both what you smell and how it tastes. Include as many descriptors as possible to define each wine, even if some may be a stretch. Here is a bit of vocabulary to choose from:
   - **Fruity descriptors:** Apple, apricot, crisp, cucumber, fig, fruit salad, golden delicious, gooseberry, grapefruit, green apple, honeydew melon, kiwi, lemon, lime, mandarin, mango, melon, nectarine, peach, pineapple, prune, quince, rockmelon, spicy, straw, tangerine, tropical fruit
   - **Woody descriptors:** Cedar, charred wood, cigar box, coffee, green wood, oaky, pencil shavings, resinous, sandalwood, sawdust, smoky, toasty, tree bark, vanilla
   - **Barrel flavor (malolactic fermentation) descriptors:** Almond, banana, burnt caramel, butter, buttered popcorn, butterscotch, cashew, coconut, lanolin, vegemite
   - **Lees flavor descriptors:** Baked bread, creamy, cheesy, leesy, nutty, yeasty, yoghurt
2. As you go back and forth from one wine to the next a few times, be sure to cleanse your palate with a bit of bread or plain crackers. Sip some water as well, if you absolutely must. But don't do any other food—it will easily alter the results of your descriptions. Don't worry: you can finish the bottles with dinner later.
3. Stop and take a break. Assess your descriptions. Did you use

a lot of words from a single set? That is, did you mostly use the 'fruity' descriptor list for one wine, and maybe more 'barrel flavor' descriptors for another, or perhaps you pulled equally from two different sets...or maybe your descriptions were from all the lists for one bottle, but not the other?

4.   Now throw the bottles into the fridge for 30 to 45 minutes, until it has a good chill on it. Repeat the tasting exercise. Take good notes. Has anything changed? Do either of the wines seem more fruity or more toasty or more creamy? Again, assess your lists of descriptors for each wine. How are you generally characterizing them? *How can two wines made from the same grape coming from the same region be so different?*

What to look for:

You probably found that the un-oaked Chardonnay displayed mostly, or exclusively, fruit flavors. Did you use any other descriptors to define the Jewel Chardonnay? However, that oaked Chardonnay may have had descriptors from across the board. Did you use any fruit flavors to describe the oaked Toasted Head? Or did you pull mostly from the woody list?...or the barrel list?...or maybe all the lists?

Hopefully you discovered that the Jewel was pretty simple and straightforward, while Toasted Head had a lot more complicated smells and flavors happening in it. So, what's going on here?

Why this is:

Several different things are going on here that make the smells and tastes of these two Chardonnays radically different. It all has to do with winemaker influence. For all I know, these guys could have been using the exact same grapes from the exact same vineyard (likely not though), so these radical differences in style are created by how different folks make different decisions in the winemaking process to create radically different styles. Let's take these things one at a time.

Stainless Steel: the Anti-Oak

If a winemakers decides to do the initial fermentation, any secondary fermentation, and finished wine storage without the use of barrels—oak barrels to be exact—then the finished wine will express only the fruity characteristics of the grape varietal(s) used, in this case Chardonnay. Stainless steel tanks impart zero flavor and zero aromas to the wine. That's the easy one. So how about...

The Use of Oak Barrels

The alternative to stainless steel is using barrels, either as a container in which to do the initial fermentation process (called barrel fermented; BF), a secondary fermentation process (usually malo-lactic fermentation; MF), or as a place to age the finished wine after all fermentation processes are complete (barrel aging). These different processes each contribute unique aromas and flavors to the finished wine, and we'll get to those in a minute, but first...

Why use oak at all?

Oak is the preferred wood for making the barrels and casks in which wine is aged. The wood is slightly porous, which creates an environment ideal for interaction with the fluid wine to draw out some things from the barrel. What things? Oak imparts flavors, colors, and tannins which are desirable for most red wine as well as some white wines. Tannins increase the body and complexity of the wine, soften the acidity of the wine, and increase the ability to age the wine, as well as helping to integrating its flavors together. Oak is the Prozac of wine: it tempers its extreme elements and deadens its brilliance.

Extraction of color from the wood may bring darker yellow hues to white wines, although impacts to red wines are negligible. And wood imparts a cacophony of flavors to the wines, dependent on which processes the winemakers uses the wood for, as we'll see below. Before we go further, let me point out a variety of other factors tied to the wood itself which can produce radical differences in the flavors and aromas of finished wine:

❖ *French vs American oak:* The two main types of oak used to make barrels in the entire world. Just know this: French oak is usually considered more subtle and lighter in flavors (this is due to it having a tighter grain than American oak, which means the wine doesn't 'get in' to it as much), while American oak is much more flavor-imparting (looser grain) and produces flavors sometimes referred to as aggressive, raw, or extreme. Maybe we'll try an experiment later to demonstrate this.

❖ *Age of barrel; new v. used.* Brand new oak barrels, as you might expect, pack the biggest punch for smells and flavors because they haven't had anything extracted out of them yet. Many wine labels will brag about the use of new wood—"Aged two years in new French oak." Older, used barrels that have already been thru the process a time or two have less to offer. But barrels are expensive (like $800 a pop nowadays), and so winemakers want to get the greatest use out of them as possible. They will also use the age of the barrel intentionally to impart different flavors at different times. You might see a wine description that says: "six months in new oak, two years in used barrels." Just know this: new oak=more powerful flavors, used oak=less so, more subtle flavors.

❖ *How long you barrel age.* As referenced above, how long you let the wine hang out in the barrel affects the strengths of the aroma and flavor that is imparted. Two years of aging in oak may give twice the impact that just one year in oak does (its probably not that exact, but you get the idea.)

❖ *How toasted the barrel.* Finally, there are ways to get even more bang for your buck out of a barrel. One way is to toast, or char, the inside of the barrel before you use it. This allows even more access for the wine to get into the grain of the wood. A light toasting happens naturally as part of the barrel making process, but you can go out of your way to burn it up even more, thus allowing more access to those wood flavors. You just had a wine called 'Toasted Head'; and now you know what they are referring to. They burnt up the barrel ends (the top and bottom of the barrel also known as 'the heads') in order to extract more flavors and aromas. As a final note, after new barrels are used for a couple years or so, it is customary to re-toast them again in order to 'freshen them up' a bit to regain access to the tannins and flavors in the next layer of wood.

As you may have tasted already, the use of oak imparts flavors from the 'woody' descriptor list, namely the smell and taste of freshly sawn oak, a toasty or smoky smell or flavor, and one of great significance to modern wine taste: vanilla. Vanillin is a chemical component in the wood itself, and even in trace amounts can leave a whiff of vanilla and bring a perception of sweetness to the finished wine. But to be sure, the addition of those flavors is NOT the primary reason to use oak. Its really all about the acids. First and foremost in this discussion is the softening of acids…

Acids in my wine? What? I don't want to get burned!

Acids for your consideration: Tartaric & Malic & Lactic
The tart taste of dry table wine is produced by the total quantity and the kinds of acids present. Tartaric and malic are the major wine acids. These two acids are present when the grapes are picked, and they are carried over through the fermentation process into the finished wine.

- *Tartaric* Few fruits other than grapes contain significant amounts of tartaric acid. One half to two thirds of the acid content of ripe grapes is tartaric acid, and it is the strongest of the grape acids. Tartaric acid is responsible for much of the tart taste of wine, the 'pucker factor', and it contributes to both the biological stability and the longevity of wine. How? It is resistant to decomposition and affects the PH of the wine to discourage bacterial growth. The amount of tartaric acid in grapes remains practically constant throughout the ripening period. However, the situation in wine is different. The quantity of tartaric acid slowly decreases in wine by small amounts over time.
- *Malic* Malic acid is prevalent in many types of fruit. This acid is responsible for the tart taste of green apples. Malic acid is one of the biologically fragile wine acids, and it is easily metabolized by several different types of wine bacteria. Unlike tartaric acid, the malic acid content of grapes decreases throughout the ripening process, and grapes that are grown in hot climates contain little malic acid by harvest time. Grapes grown in cool regions often contain too much acid. High acidity results in excessively tart wines, so the winemaker has a problem. Malolactic fermentation (ML) can reduce wine acidity. When wine goes through malolactic fermentation, bacteria convert the malic acid into lactic acid. Lactic acid is milder than malic acid,

and ML fermentation is a standard procedure used to reduce the acidity of wines made from grapes grown in cool regions. More on this below...

- *How climate affects this:* Tartaric and malic acids are produced by the grape as it develops. In warm climates, these acids are lost through the biochemical process of respiration. Therefore, grapes grown in warmer climates have lower acidity than grapes grown in cooler climates. Sugar production is the complete opposite of acid production. The warmer the climate, the higher the sugar content of the grapes. In summary, warmer climates result in high sugar and low acid whereas cooler climates result in low sugar and high acid.

So what's this malolactic fermentation stuff?

The malolactic fermentation (ML) is an important natural process for adjusting acidity. The ML lowers the acidity by converting malic acid to lactic acid and carbon dioxide. Many white wines are encouraged by the winemaker to undergo ML and almost all red wines "automatically" undergo ML. In some grapes the acid is so high [example: Chablis] that they require a malolactic fermentation to lower the acidity. Since some wines have less malic acid in them than others, the ML is not as significant in shaping the wines as in those with a higher malic acid content. For example, a white Burgundy (which are 100% Chardonnay) typically contains less malic acid than a Napa Valley Chardonnay. Therefore, when a white Burgundy undergoes ML, very little acidity is lost and the character of the wine is preserved. On the other hand, a California Chardonnay contains more malic acid so when it changes to lactic acid the acidity can change appreciably.

And here's the result on your finished wine: wines that undergo malolactic fermentation have the buttery, creamy, banana flavors from your list above. That's the lactic acid part.

And finally, what the hell are lees?

Lees are the bi-products of the fermentation process; it's the junk at the bottom of the barrel or the tank. They are mostly comprised of the dead yeast which fall out of suspension after they have completed their heavenly, alcohol-converting work and have passed on to a better place. Lees Aging, or Aging *SUR LIES* refers to aging a wine on its post-fermentation sediment (lees) in order to enhance its texture, longevity and flavor complexity. Flavors developed may include baked apple, baked bread, hazelnut, honey and aged cheese. Lees stirring (sticking a goofy looking spoon into the barrel and whipping up the lees) is called "*bâtonnage*". By the way, very, very few wines undergo heavy lees aging, and the wine label will almost always prominently display the words "*sur lies*" on it so you know what you are getting into.

In summary:

- ❖ The grape variety gives you the fruity flavors/aromas, which change from variety to variety

❖ The oak barrels give you the woody and vanilla flavors/aromas
❖ The malolactic fermentation gives you the buttery flavors/aroma
❖ And the lees/sur lies gives you the yeasty, bready, cheesy flavors/
   aromas.

What process or combination of processes the winemaker decides to use contributes to what flavors and aromas are in the finished wine. Remember, not all of these things happen to every wine. Every wine has its varietal characteristics to be sure, but not every wine has the other stuff— but a lot do. Most reds do have the malolactic component, which you can detect in the rounder, vanilla or butter hints at the end of a big Shiraz or Cab. Many whites go thru it too, as can be tasted in big 'oaky buttery' California Chardonnays. Most wines also go thru some time in oak, but you now know how different variables affect those flavors/aromas as well. And just some stuff goes sur lies…and they are pretty easy to pick out: if you smell cheese *in* your wine, as opposed to having cheese *with* your wine, than its probably had some time on the lees.

If all that doesn't confuse you, then someone come explain it back to me….

	Un-oaked Chardonnay	Oaked Chardonnay
Just opened		
Color:		
Bouquet (smell):		
Taste:		
Body (heavy/light):		
Other comments:		
Chill - One hour later		
Bouquet (smell):		
Taste:		
Body (heavy/light):		
Other comments		
With any food		
Taste:		
Other comments:		

Lesson 5: Dry vs. Sweet: A Sugar Showdown

What it means; what it tastes like. How sugars and acids in wine interact to create different styles. How it the winemaker manipulates the process to achieve these affects. What are the processes involved.

Dry and sweet are the opposite ends of the wine spectrum. Describing a wine as dry means that the yeasts have fermented out all of the sugar during the wine-making process, leaving a wine with no residual sugars, usually a higher alcohol content, and a bit of tartness. Sweet and semi-sweet wines are those that taste sweet to our palate because there is some residual sugar in the wine, or a sweetening agent has been added at some point in the process. Of course, there is a whole spectrum of flavors in between the extremes of 'completely dry' to 'sickeningly sweet' and where a wine ends up depends on the winemaker's craft. For this exercise we will be using Riesling because it is probably the varietal most widely used to make the greatest range of sweetness levels, and it has great geographic diversity too—almost all wine-producing countries produce a Riesling or two…or ten…or a thousand.

What to grab:
- One bottle from each list of the following from your local wine shop or grocer:

  Dry Rieslings

  Bottles $5-$15
  - ➢ Milton Park Riesling (Australia)
  - ➢ Stonehaven Winemaker's Selection Riesling (Australia)

  Bottles $16-$30
  - ➢ Trimbach Riesling (France)
  - ➢ Hogue Cellars Genesis (Washington)
  - ➢ Trefethen (California)
  - ➢ Hugel Riesling (France)
  - ➢ Domaine Zind-Humbrecht (France)

  Bottles $30+
  - ➢ Kilikanoon Mort's Reserve (Australia)
  - ➢ Domaine Weinbach StoneCuvee St. Catherine (France)

  Sweet Rieslings

  Bottles $5-$15
  - ➢ Nalbach Liebfraumilch (just slightly sweet & contains Sylvaner) (Germany)
  - ➢ FunF (Germany)
  - ➢ Dr. Loosen Dr. L Riesling (Germany)
  - ➢ Selbach-Oster Spatlese (Germany)
  - ➢ St. Urbans-hof Riesling (Germany)

  Bottles $16-$30
  - ➢ Weingut Maxford Richter Brauneberger Juffer-Sonnenuhr (Germany)
  - ➢ Schafer Dorsheimer Goldlock Riesling Spatlese (Germany)

> **Bottles $30+**
>   ➢ Weingut Maxford Richter Mülheimer Helenenkloster Riesling
>   Spatlese 1989 (Germany)
>   ➢ Joh Joh Prum Graacher Himmelreich Auslese (Germany)
- Bread or unsalted crackers. Plain, with no fancy flavors attached.
- Water

Don't worry if you can't find these exact labels. You can always ask your local wine store salesperson to help you pick out a good representation of the style I have described above. Show them the list and tell them to get you something close to it.

What to do:

1. First, before you even chill it, open both bottles immediately and pour a glass of each. Don't drink it yet! Just set those glasses on the table for now, and it might be wise to label each glass accordingly (sweet and dry). Take the rest of both bottles and stick them in the fridge for 45-60 minutes. Once they have received a good chill, you are ready to go. Take them out and pour two more glasses, one of each of course. Now you should have 4 glasses on the table: 1 warm dry, 1 warm sweet, 1 cold dry, and 1 cold sweet.

2. Start with the ***chilled*** wines: Assess the chilled dry Riesling first for color, aroma and taste. Record all impressions. Then do the same for the chilled sweet Riesling. You will easily detect a huge difference in flavors…and perhaps even different aromas as well. Include as many descriptors as possible to define each wine, even if some may be a stretch. Here is a bit of vocabulary to choose from:
    - **Riesling descriptors:** Apple, apricot, citrus, honey, honeydew, honeysuckle, lemon, lime, peach, tropical fruit, rose petal, flower, floral, berry, diesel, lanolin, petrol, kerosene, gunmetal, geranium, rose, orange blossom, jasmine, grapefruit.

3. Now let's have some fun with the ***warm*** wines. **Pay attention to this: DON'T compare the warm ones to each other! Compare them side by side with their cool counterpart.** In other words: Try the warm dry Riesling, then do the cool dry Riesling. Next, try the warm sweet Riesling, then the cool sweet Riesling. Record all impressions. Hopefully you will detect a difference in the perception of sweetness/dryness as you go back and forth between these two sets. Be sure to taste the dry before sweet (always a good idea in any exercise) and if you want to go back to reassess the dry ones again, be sure to cleanse your palate as much as possible with bread and water. Sweetness can 'hang out' on your palate very easily, and if you go straight back to a dry wine after just having a sweet one, it creates the perception of outrageous dryness in the mouth.

4. As cited above, as you go back and forth from one wine to the next a few times, be sure to cleanse your palate with a bit of bread or plain crackers. Sip some water as well, if you absolutely must. But don't do any other food—it will easily alter the results of your descriptions. Don't worry: you can finish the bottles with dinner

later.

5. So what happened? Could you pick out some of the same fruit hints from both bottles, regardless of temperature? And obviously you picked out some taste differences. Right? But what changed with change in temperatures, for both wines? What's going on here?

What to look for:

The obvious: the dry wine makes you pucker a little bit, the sweet one tastes...um....sweet! However, since both wines are made from the same grape, you should smell some similarities, and if you concentrate on your palate you can even taste some similarities—but you have to work hard to see past the sweet/dry differences in order to pick up those varietal similarities on the palate. I don't expect you to pull it off, but just think about. It is the same grape in two very different wines.

The other thing to look for is that chilling the wines should decrease the perception of sweetness. The chilled dry will seem even drier than the warm, and even more pronounced should be that the warm sweet will seem even sweeter than its well-chilled version. Did it work?

Why this is:

Two different things are going on here that make the smells and tastes of these two Rieslings radically different. One is the winemaker influence. A winemaker could have easily made 10 different wines with ten different levels of sweetness from grapes in the exact same vineyard. The radical differences in style are created by how different folks make different decisions in the winemaking process to create radically different styles. The second thing has to do with temperature and its affect on wine. Let's take these things one at a time.

Winemaker influence on sweetness level

As you already know, sugar accumulates in the grape during the ripening process. And hopefully you remember that climate and weather during the growing season affects this process: cool climates with shorter growing seasons have a harder time ripening fruit and getting higher concentrations of sugar; like in Germany. And a bad growing season can affect sugar concentration even in a warmer climate. Heavy rains during harvest can water-log the fruit, causing a decrease in percentage of sugars, which happens in France all the time. You also know that the amount of sugar in the grape is going to affect the amount of alcohol that can be produced, as well as the overall balance of the wine. And now as you have tasted, how much sugar they leave in the finished wine radically affects its taste.

The winemaker considers all of those factors, the market demands, and his own stylistic desires during the wine production process. For sure, they all need a certain level of grape sugars to produce some alcohol. And given climatic/weather variability, they may need to manipulate sugars in order to 'fix' some problems in the wine. But mostly how sweet a wine ends up being is all about what style of wine the producer is striving for.

The winemaker can affect the amount of sugar, and therefore amount of sweetness, in the wine in four distinct ways:

1. **Over-ripening of the grapes:** The winemaker/viticulturalist (and these guys always work in tandem—and sometimes it's the same guy!) constantly assess the grapes during ripening in order to figure out precisely when they have the desirable sugar/acid balance to craft their wine. They can choose to allow the grapes to stay on the vine as long as possible in order to maximize the grape sugars accumulating. The alternative of course is that they could choose to harvest early thereby getting less sugars, but that's kind of rare. In some cases, they allow the grapes to stay on the vine so long, that they begin to shrivel (basically becoming raisins) which concentrates the sugars even further. In German Eiswein, they allow the grapes to actually freeze on the vine, which locks up the water in the berry, thus concentrating the sugars in a different way. All of these processes increase fruit sugar BEFORE the wine making process begins.

2. **Stopping the fermentation:** After the harvest and pressing and during the fermentation process, the winemaker has the ultimate responsibility of how far to let that process go. You know the deal: the yeast eat the sugars and spit out alcohol and they will keep doing their work until they run out of food or the alcohol level gets too high for them to survive. But the winemaker can stop the yeasts' work anytime he/she desires. They could let the juice ferment for a day, a week, a month or a year if they want. If they stop the fermentation before the yeasts run out of food, then there will be some residual sugar left in the wine, which you will taste as sweetness. If they let the yeast eat all the sugar, then there is no residual sugar, and the wine will taste dry. No sugar, no sweet! They can achieve this work-stoppage by several means: 1)dropping the temperature of the wine, which makes the yeast drop out of suspension 2)filtering the wine to get the yeasts out and/or 3) dumping a bunch of higher proof alcohol into the batch, which kills the yeast (that's how Port wine is made). All of these processes maintain wine sugar DURING the wine making process.

3. **Adding in unfermented grape juice:** The winemaker can also manipulate the amount of sugar in the finished wine after the yeasts have done all their work. Perhaps it was a rough year and you have low fruit sugars which are going to equate to a low alcohol wine--but you want to the wine to taste sweet. If you stop the fermentation early in order to preserve some residual sugars, then the wine will have unacceptable low alcohol. If you let the fermentation go, you get higher alcohol, but it will be

dry. What to do, what to do? The Germans are often faced with this conundrum so they created a process named süssreserve in which you hold a portion of the freshly-pressed grape juice aside from the fermentation process, and then after the wine is finished fermenting (and therefore 'dry') you dump the unfermented sweet grape juice back in. Presto! You got your alcohol, plus some sugars in there to make it taste sweet. This process increases wine sugar AFTER the wine making process ends, or perhaps its better to say after the fermentation stage is complete since blending is part of the wine-making process. If you did a sweet German Riesling for this exercise it probably was by this method. As you can imagine, throwing in some of the unprocessed juice adds a lot of varietal flavors back in too.

4. **Adding in sugar:** This last one is a catch-all category that is done for a variety of reasons. The appropriate term is chapitalization, and it simply means adding in a bunch of sugar before or during the fermentation process. Why do it? To increase alcohol levels, to increase sweetness levels, and/or to balance acidity. We already discussed sugar levels and alcohol production. But in bad harvest years when the grape sugars are too low, winemakers have the additional burden of having the acids too high. Remember, acids and sugars do a balancing act, so as one goes down the other will be too high. Adding sugar to the **must** (pressed grape juice) before fermentation often occurs to regain balance in highly acidic situations, and/or because they don't have enough sugars to produce adequate levels of alcohol.

All of these processes and decisions are part of the winemaker's arsenal to affect the sweetness/dryness level of the finished wine. Not everyone uses all of these tools. The Germans and French often have to manipulate the sugar levels to achieve their end products, while a place like California has such a good climate with little variability that such processes are rarely required.

Temperature affecting perceptions
Simply put, chilling wine only really does one main thing: it decreases the perception of sweetness. Not the actual level of sweetness mind you—that remains the same. But how sweet it tastes to you will decrease significantly as you get the wine colder and colder. This is why we typically chill white wines, as they usually have at least a modicum of residual sugars, and why we typically don't chill reds as they are usually completely dry. Chilling the whites allows the fruity notes and flavors to appear at the forefront of your palate instead of the sweetness which (as you may have experienced) tends to overwhelm the senses. Red wines, being dried out, do not benefit from the temperature drop. You don't have to chill anything if you don't want to—that's all personal preference. Try it with other wines to see what you like; drink a Chardonnay warm and then cold, or a Sauvignon Blanc. See how temperature affects their flavors. Maybe you like it more served warm?

One final note: People often confuse fruitiness with sweetness. You will experience this first hand in these class exercises here very soon and I want you to be aware. In some varieties like Riesling, Gewürztraminer, and Viognier, the aromas often provided by the fruit are so sweet and so honeyed and so floral and so luscious…it just makes your mouth water! But they might be totally dry wines! We have been programmed to associate smell with taste so much that if we smell honey, we expect to taste something sweet…like honey. But you now know that the winemaker can craft the wine to have those floral hints, but be bone dry. Depending on what dry Riesling you used for this exercise, you may have experienced this already. So try to refrain from describing a wine's nose as 'sweet', because you can't really smell 'sweet'—that's a flavor! When you smell 'sweet' you are actually getting fruit characteristics that I want you to associate with those grape varietals, not with sugar.

Summary:

* Dry versus sweet are opposite ends of the wine spectrum, and there is tremendous variation in between
* The winemakers can manipulate the production process to achieve these ends in a variety of ways
* The grape variety gives you the fruity flavors/aromas, which don't necessarily change with differences in sweetness levels
* Changing the temperature of a wine decreases the perception of sweetness: essentially, chilling a wine makes sweet wines seem less sweet, and dry wines more dry

	Dry Riesling	Sweet Riesling
Chilled		
Color:		
Bouquet (smell):		
Taste:		
Body (heavy, light):		
Other comments:		
Warm		
Bouquet (smell):		
Taste:		
Body (heavy, light):		
Other comments:		

Lesson 6: Color me Clueless: is there anything to a wine's color?

Why wines come in a variety of colors. How a winemaker achieves this variety. What the wine's color is really an indicator of. Should you really care about the color of the wine. Or is it all just a pigment of your imagination?

What to grab:

This isn't really a cheapie experiment if you go for the gusto with the Zinfandels. Feel free to work with others, but it is essential for this lesson to get *one* of the *sets* of wines listed below. If you can't find the exact wines listed below, just ask the folks at the wine store to hook you up with something close. Show them this list and they will know what you're talking about.

- Grab one of these 'sets' of wines from your local wine shop or grocer:
    - any light easy Chardonnay $7-$11
    - any heavier, barrel-aged Chardonnay $8-$20
        OR
    - Tres Ojos Rouge (Spain) $8
    - Tres Ojos Rosé (Spain) $7
        OR
    - Domaine Grand Veneur Cote du Rhone (France) $9.95
    - any Tavel rosé (France) $11.95
        OR
    - any White Zinfandel (California) $5-$8
    - any Blush/Rosé Zinfandel (California) $6-$9
    - any Red Zinfandel (California) $16-$36
- A hard Italian or Spanish cheese, like Parmesan or Asiago or Peccerino
- Bread or unsalted crackers. Plain, with no fancy flavors attached.
- Water

Don't worry if you can't find these exact labels. You can always ask your local wine store salesperson to help you pick out a good representation of the style I have described above. Show them the list and tell them to get you something close to it.

What to do:

1. This is really the only exercise where focusing on color is of primary importance. So the first thing to do after opening the bottles in your set is to pour a glass of each and set them up in front of you. Contemplate and describe the colors you see, with a particular eye towards describing the differences in intensity or richness of the color between your two (or three) wines. It may help to be conducting this session on a white table cloth, or at least lay down some white paper under your wines

so you can really see the colors in their purest light. Pick up the glass and tilt it at different angles as you do the assessment to check out the color intensity at the 'edges' of the liquid.

2. After your thorough color check, go ahead and do your standard sniffing and drinking assessment. ***For order of wines: Always do the lighter yellow before the darker yellow with whites; the rosé before the dark red; and if you are doing the Zinfandel line-up, do the rosé, then red, then the white last.*** What is the aroma, the body, and the taste of each wine? Record all impressions. Write down some descriptions of your experience, both what you smell and how it tastes.

3. Also for this lesson, try to describe how the wine *feels* in your mouth. Is one wine heavier or lighter than the other? Is one wine thicker or thinner than the other? As always, include as many descriptors as possible to define the wine, even if some may be a stretch. Be sure to take a short break, drink some water, and hit up some crackers in between each of the wines. You got to give your palate a cleansing period in order to adequately assess the mouth feel of these different wines.

4. And that's really it. This is a pretty short lesson, as it's supposed to be mostly on the color of the wine, which only really takes five seconds to determine. However, I hope you established some connections between the color and the feel of the wine in your mouth. We'll come back to this idea in the next lesson, but for now keep the rest of the bottles for dinner. Try with a variety of different foods. What works? What doesn't? What foods bring out more flavors in the wine? What foods overwhelm the wine? I want you to experiment as much as you want to see what works best for you.

What to look for:

You may have loved or hated the wines you just drank. But let me assure you that the color of the wine had nothing to do with that opinion...unless you just really groove to the hues, and therefore liked the rosé the best because pink is your favorite color. The point I'm trying to make is that color is absolutely no indicator of a wine's quality. [With one obvious exception: a white wine that looks brown in color is complete crap.] There are good red wines and bad red wines just as there are good pink wines and bad pink wines. And there are a million variations of the white and yellow and red color spectrums in wines, so there really is no 'correct' color for a wine to be.

I personally think the 'assessment' of a wine's pigments is primarily poppycock! At least it is when wine writers and wine snobs try to equate the color with some level of quality.

However, color can be considered an indicator of a wine's *style* and even of its *body*. And that is a horse of a different color altogether. Wow! What a play on words! After this exercise, I hope you now know the stylistic difference between a red Zin and a white Zin...because that is like night and day. It's the same grape, but the difference in color makes for two radically different styles of wine. For white wines, a darker yellow hue typically means its been barrel aged longer, which means more oak flavors.

In terms of a wine's *body*, I hope you paid attention enough to 'feel' the difference between the wines as well. The French or Spanish reds should have felt 'heavier' or 'thicker' in your mouth than the rosés did. If you did the whites, and are really good at this, you may have detected that the darker yellow Chardonnay felt 'richer' or 'fuller' when you swished it around your palate than the lighter ones. Well did it? And possibly the red Zin was much fuller-bodied than the pink Zin, unless the pink was an ultra-sweet version. So why is it that more color or richer color in a wine somehow equates to more body. And now that I think about it, how the hell can we have three different colored Zinfandels that are all made from the exact same grape?

Why this is:
The answers to these questions of color all derive from the skin of the grape...which is fitting, since all the colors in wine are also derived from the skin of the grape. What do I mean by this double speak? Just this:

All grape juice is clear. Yes, all of it. I know you probably drank lots of Welch's grape juice as a kid, and that stuff is dark, dark purple, so you have assumed that it must have been made from dark purple grapes which when squeezed produce dark purple juice. But your assumption is fundamentally flawed. There are green grapes and red grapes and black grapes and every shade of grape in between, but here is the real deal: when pressed out of the grape, all grape juice runs clear. Test it out the next time you are in the grocery store. Pick as many different colored grapes as you can find and squash them on the counter. What color is the liquid that comes out?

So how do we get red wine if all the juice is clear? Via a winemaker tool called **skin contact**. All the pigment of a grape is locked up in its skin, so if a winemaker wants to make a wine with color, he/she must allow the freshly squeezed juice to hang out with the skins for a while. In the winemaking process, the grapes are picked, brought into the winery, and the juice pressed out. At this point if the winemaker wants color in the wine, he will keep the juice and smashed skins together in one big pot... thus allowing the juice to have *contact* with the colored *skins*.

Over time, the pigment is 'washed out' of the skin and blends into the juice solution. How long the winemaker allows that skin contact to occur equates to how much pigment ends up in the juice. Short skin contact = less color; longer skin contact = more color; no skin contact = no color! Hey! This is easy to understand! Some wines have minimal skin contact and only pick up a bit of bit of color, while other wines may go through the whole fermentation process with the grape skins hanging out in the vat... which of course produces lots of dark color extraction.

Now the color of the grape skin has a lot to do with this too. Green grapes have little to no pigment, and produce white wines (unless they are used in blending with other darker grapes). Chardonnay grapes are green, so it is pointless to waste time with any skin contact; they just produce a white wine. Really dark red and black grapes have lots of pigment, and thus are used to make red wines. Cabernet Sauvignon and Syrah are very dark grapes, and can produce really dark wines. And there are some grapes, like the Pinot Noir, which are somewhere in between. They have thin, lightly pigmented red skins

and even extended skin contact will only yield so much color from them... which is why a lot of wines made from the Pinot don't have tremendous amounts of color in them.

But don't get confused by that anomaly! The primary reason a wine is light red or dark red relates to how long the skin contact occurs. If a winemaker is making a pink rosé, the skin contact is minimal—maybe only a few hours. However if the winemaker is going for a big, deep ruby red wine, then of course the skin contact will be as long as possible. And yes, this is why you can produce a white Zin and a pink Zin and a red Zin all from the exact same batch of juice. You dig? It's all about the length of skin contact!

But how are color and wine body related then?
Let's combine the points made above to point out a few things you can tell about a wine from its color:

Color is not the only thing that gets extracted out of the skins. There is an additional process that is occurring during this skin contact process that I haven't told you about yet. So here we go. Stuff called *tannins* are also present in high amounts in the skins, stems, leaves, and seeds of grapes. Just as more and more color is extracted in longer and longer skin contact sessions, so it goes with the tannins. Thus...as the juice picks up more color, it is simultaneously picking up more tannic character.

I've referred to tannin in several of these lessons, with more to come later, but let me encapsulate a few of the main points about it again here. Tannins give the wine the mouth-pucker feel, the mouth-drying sensation, and perhaps even a harsh abrasive nature when they are present in high amounts. They are primarily found in red wines, for reasons that are perhaps now obvious to you. What? Not so obvious to you? Well here's the answer: it's because white wines don't go through skin contact phase and therefore have not only no color, but no tannins either.

But hold on! That's not to suggest that white wines can never have any tannin, because they can. Tannin is also found in wood...and in barrels made out of oak wood to be more precise. So whenever a winemaker ages his white wine in oak barrels, he/she is imparting some tannin, and even a little bit of color, into the final product. But wait a minute; I thought we were talking about how color can be an indicator of a wine's body?

Oh yes, back to the point. Any red wine with intense colors by default has a high tannin component, which adds more to the body of the wine, making it feel heavier, thicker, fuller, or simply giving it more mouth feel (more to chew on.) Since really dark red wines must have had a long skin contact period to get all that pigment, we can assume that they have a high tannic component as well. This would mean that the wine probably has a lot of body. We *cannot* however assume that the wine is very harsh or astringent because of high tannic presence. The winemaker could have balanced it out with other elements to make it smooth—try a big, dark, yet smooth Australian Shiraz sometime, which displays a 'New World style'.

Having said that, a dark red wine could very well be a big mouth puckering bomb when young, if it is crafted in the 'Old World style'. The French or Spanish dark reds should have felt 'heavier' or 'thicker' in your mouth than the rosés did. If you detected that difference, then what you have experienced is how increased tannins 'weigh' down the body more. Either way, we can predict that very dark wines will have a lot of body or mouth feel due to the tannin component.

And this same reasoning applies to white wines, albeit at a much lesser scale: darker yellow hues would indicate some barrel aging, which would tip you off to some tannic structure, and therefore a slightly heavier body. If you did the Chardonnays, did the darker colored one seem chewier or thicker in your mouth than the light colored one?

But all this is quite academic. Point is, color can be an indicator of a body... but not of any level of quality.

So how is color related to style?
To finish, just let me tip you off to a few other specific things that a wine's color will tell you about certain wines:
- As suggested in the section above, if it's a black as ink colored wine, it will likely have a big heavy body. The lighter the color, typically the lighter the body.
- If it's a red or black grape that typically makes red wine, but you see it as a white wine, you can safely bet that the style is sweet to sickening sweet. White Zinfandel your best example here, but I have even seen 'White Merlot', which is an overt marketing ploy to cater to non-wine drinkers who just like something as close to Kool-Aid as they can get. But who am I to judge? If you like it, then drink it!
- Pink wines from European countries like France are typically bone dry wines that are delightful for picnics on the Mediterranean coast.

We call them rosé wines.
- Pink wines from California are often semi-sweet to full on sweet and are best enjoyed with cotton candy at a seaside carnival at Venice Beach. We call them blush wines.
- White wines from the 'New World' with dark yellow hues have often seen barrel time and a malolactic fermentation, and thus would have more oaky, buttery flavors (mostly California Chardonnays).

Once again, color can be an indicator of style…but not of any level of quality of that style.

Summary

As all grape juice is initially clear, the color of the finished wine is entirely dependant on the desires of the winemaker. They can make any grape juice into a clear, light pink, or dark red wine if they so choose. Manipulating the time of *skin contact* is what gives the winemaker the versatility to produce these different colors of wine. A little trivia to make the point: 2 of the 3 grapes used in Champagne production are dark-skinned, but all the finished wine is light and bright in color!

The grapes' color also plays a role in this: you can't make a red wine out of a green grape because it has no pigments to transfer to the liquid. The darker the grape berry, the darker the wine can *possibly* be, if a lot of skin contact occurs.

Besides color, red grape skins also contain high levels of tannin which can be imparted to the wine. Tannins increase a wine's longevity and its body or mouth feel. A wine's color can be an indicator of that wine's body: typically, the more color, the heavier the body.

And finally, color is not really an indicator of wine quality. Feel free to not pay attention to color as often as you like. I usually ignore it altogether. But it's good to know how this stuff works, right? At least now you can bust some snob's chops at a dinner party if he starts trying to jive you about how he can tell that his Pinot Noir is so perfect because of its color. Yeah, right.

	Wine 1	Wine 2	Wine 3
Color:			
Bouquet (smell):			
Taste:			
With Cheese:			
Body (heavy, light):			
Other comments:			

Lesson 7: Rock Your Body but Keep Your Balance!

Describing light, medium, and full-bodies wines. What these terms mean. The major factors which contribute to wine body. How those same factors play off each other to affect a wine's balance. The processes involved, and the winemaker's role in affecting body and balance.

Who doesn't love a nice, balanced, full-figured body? Whoa, hold on there…I'm talking about in wine of course! As you start to drink wines, you should pay attention to how it feels in your mouth. [insert your own joke here] Seriously! You will hear folks describing a wine as 'insipidly thin' or with 'good mouth feel' or even as 'full-bodied.' What does all of this mean? And is it merely descriptive, or are the *body* and *balance* of a wine somehow an indicator of wine quality? Well, let's find out.

I'm going to combine a whole bunch of information together into this one simple lesson which will attempt to relate these things together so that you, the novice wine drinker, will at least understand the basics of a wine's **body** and **balance**…and also so you can hang in a wine snob conversation that may allude to a lot of these factors that I'm going to lump together.

For starters, when people refer to a wine's **body**, they are describing how thick or thin or how oily or watery it feels in the mouth. You are going to taste a bunch of different wines with a bunch of different levels of *body*: some will seem obviously light, others extremely heavy, with most falling somewhere in between. Some of these wines have one particular trait which gives them some *body*; other wines have a combination of traits which perhaps gives them much more *body*; and of course one of these beverages has no *body* at all. Poor thing. Doesn't everybody need some *body* sometime?

I have picked certain wine styles that have distinct body characteristics, however the wines' *balance* will have to be determined by you, the drinker, as that is specific to each particular wine, its vintage, and its craftsmanship. **Balance** refers to the interaction or harmony between the major components of the wine.

What are these wine components that work in tandem to create the balance, and the components that add body to the wine? I'm glad you asked! For our lesson, the wine components responsible for both of these concepts are: sugar, acid, tannin, and alcohol.

Be forewarned: lots of wine makers and wine writers and wine retailers and wine gurus of every ilk would disagree with this lesson, arguing that I'm oversimplifying this complex topic and mixing terms which don't belong together. They will show you statistical charts and Venn diagrams and pictures of a human tongue divided up into taste regions in order to explain all the nuances of describing wine. But I say: the hell with them. Let them write their own damn book. I'm telling you what you need to know so that you can drink this now…and swiftly be savvy enough to speak about it to snobs of all sorts.

So let's keep our balance, and proceed to mingle with some bodies…

What to grab:
- Water, and a lot of it. At room temperature too — not cold.
- Try to get a bottle of as many of these types of wine as you can afford. If pressed for funds, just grab a 'thematic pair to compare' as outlined below. Group play is encouraged in this lesson if you want to do all the comparisons at once. Make it a scavenger hunt affair: send out eight people on a mission, each to obtain one of the following:
    - ➡For acid impacts on body:
    - ➢ Any crisp, acidic white wine like a Vinho Verde or a Sauvignon Blanc
    - ➢ Any creamy, oaky Chardonnay from California or Australia
      - ➡For sugar impacts on body:
    - ➢ Any low-alcohol, semi-sweet white wine like a Liebfraumilch (i.e. Blue Nun)
    - ➢ Any low-alcohol, really sweet white wine like a German Riesling that has the word Spatlese or Auslese on the label.
      - ➡For tannin impacts on body:
    - ➢ Any dry rosé wine, from Rhône is best. Look for Tavel.
    - ➢ Any really young, dark, red Rhône wine
      - ➡For alcohol impacts on body:
    - ➢ Any big red rounded wine, like a Shiraz or Cabernet Sauvignon
    - ➢ Any Port wine
- A clothespin
- Bread or unsalted crackers. Plain, with no fancy flavors attached.

Be sure to ask your local wine store salesperson to check your picks to ensure you are getting a good representation of the styles I have described above.

What to do:
As suggested, try to get as many examples of these different pairs as possible. If you can't get them all, that's okay, just do your best.
1. Now that you have a ton of wine, and I hope you are having a party, let's get to it. Pour yourself about an ounce or two of each wine into a glass. Line up your glasses, making sure that you keep them in order as they appear in the shopping list above. Just skip any pairs that you didn't acquire from the list.
2. Now, pour a glass of water, and put it in the #1 position. You should have the glass of water on the left side of your line-up, with the glass of Port on the right side of your line-up in the last position.

Grab the tasting sheet for this exercise, and let's go.

3. Don't bother with the swirl and the sniff for this exercise. Just put it in your mouth. Start with the water. Take a healthy sip. Roll it around in your mouth for a few seconds. You can even do the fake chewing motion to make sure you coat the inside of your mouth really good. Record all impressions of the body. Was it light, medium, heavy? Was it thick, thin, viscous, watery? Was it chewy, mouth-coating, or could you not feel anything at all? **Important: focus on the *feel*, not the taste**. Repeat for all the wines, making sure to keep them in the approximate order of the list.

4. Having trouble discerning the body feel? Let's make it even easier. Affix the clothes pin on your nose, pinching it shut. Repeat Step #3. Yes, even the water. Now you are forced to only feel the wine. Record any additional impressions you now have on each wine's body and mouth feel. Compare in pairs: meaning, compare and contrast the Vinho Verde and the Chardonnay, then the semi-sweet white to the very sweet, then the rosé to the red Rhône, etc.

5. Let's finish with an extreme: go back and do your first wine again, then jump straight to the last wine, which should be the Port. Wow. If you can't detect a body difference in those two, you might want to stop and sober up for a few hours.

6. This step is optional. You can continue to practice your sensory evaluation of wines with this exercise, even though it's not the point of this lesson. Go back to the beginning and assess each wine for its other attributes. Oh, but take off the clothespin first. What is the color, the aroma, the body, and the taste of each wine? Record all impressions. Write down some descriptions of your experience, both what you smell and how it tastes. You should be able to detect some less than subtle difference in flavors and aromas in these wines…and perhaps even different colors as well, but that is not of great significance for our exercise. Really work hard at picking out and describing differences between each of these wines. As always, include as many descriptors as possible to define each wine, even if some may be a stretch. Check in individual chapters for each type of wine to find a bit of varietal vocabulary to choose from.

7. If you want to go back and forth from one wine to the next a few times, be sure to cleanse your palate with a bit of bread or plain crackers. Sip some water as well, if you absolutely must. But don't do any other food—it will easily alter the results of your descriptions. Don't worry: you can finish the bottles with dinner later.

What to look for:

You probably found that as you progressed up this tasting list, the wines started 'light' then got 'heavier' and 'thicker' and 'richer.' What these descriptors are alluding to is the body, not the flavor, of the wine. In each set of two in this line-up, the second wine should have been slightly heavier or fuller-bodied than the first. And each progressive pair should have been heavier than the pair preceding it. Does it 'feel' heavier or lighter on your palate? Does some wine coat your mouth, while others just dissipate away very quickly? Different styles of wines have different bodies, and now you should be familiar with what these differences actually feel like.

Why this is:

Hopefully all of the wines you selected were tasty, but that is actually beside the point. What you should have detected is that some of the wines were 'thicker' than others, some of them were 'fuller' than others, or that some of them were 'heavier' than others. They may have all tasted great, despite radical differences in their bodies, and that's okay...there is no 'one size fits all' for wine body.

Some styles are lighter with less body...and that is good! While others can be big-bodied bombshells that feel like they need to be chewed in your mouth before you dare swallow...and that is good too! Most wines are somewhere in between. So what accounts for the radical range of body in the wine world?

Lots of different ingredients can give a wine body. The more you add of any single ingredient, the more body in the finished product. And of course if you add more than one ingredient, the body will also increase. Add three or four ingredients, and the wine body goes through the roof. Not only do these factors make up the body, but they play off one another to achieve a certain balance in the wine. So without further adieu, let's talk about how these 4 factors affect both the body and balance of a wine...

Acid The tart taste of dry table wine is produced by the total quantity and the kinds of acids present. Tartaric and malic are the major wine acids. These two acids are present when the grapes are picked, and they are carried over through the fermentation process into the finished wine. These acids provide that pleasant tartness that feels refreshing when you drink it. If you did a Vinho Verde, you were tasting almost nothing but acid...was it crisp, with a smart bite? That's the acid.

There is another acid that can be present in wines if the winemaker allows it, and that is lactic acid. Lactic acid actually does the opposite—making a wine seem creamier or smoother, thus taking some of the bite off the other acids. The California Chardonnay should have given you some of this...did it seem smoother than the Vinho Verde? Damn, it should have.

High levels of acid do increase body...as you should have detected when comparing the Vinho Verde to water. But too much acid can make the wine unbalanced by making it bite too hard. Ouch! That hurt!

Tannins are what give wine some flavors, but for our exercise we are mostly concerned with the astringency they add—that pucker, dry-mouth feeling you get when you drink a tannic red. A really tannic wine almost makes you thirsty right when you drink it, as it seems all the saliva has instantaneously evaporated from your mouth. The young Rhône red probably did this to you, as opposed to the Rosé which had much less tannic component. And you should have detected an obvious jump in body when you went from that lighter rosé to that tannic red. More tannins, more body.

Tannins are obtained through contact with the grape skins, stems, and seeds, as well as through contact with wood in an oak barrel. Tannins increase the body and complexity of the wine, somewhat soften the acidity of the wine, and increase the ability to age the wine, as well as help in integrating its flavors together over time. We will come back to this later.

As for how they affect a wine's balance, tannin pull it away from the sweet side. Just like the acids do. In fact, you can put acids and tannins on the same team, the team that is struggling against the sugar/alcohol team. A winemaker must be careful how much combo acids and tannins he allows in the wine. More tannic structure means he should back down on the acids, and vice-versa. Loading up on both will create a monster wine that is hard, full-bodied, and probably really harsh when young.

Sugar Well, duh! The amount of sugar in the finished wine gives it the varying sensation of sweetness. Little to no sugar left in the wine makes for a dry wine with no perceptible sweetness. Leave a bunch of sugar in the wine, and it tastes noticeably sweet. This is mostly prevalent in white wines, as most reds have little or no residual sugar in them after fermentation.

[Remember, don't confuse 'sweet smells' with sweetness on the palate! You can feel residual sugars because they increase the body of the wine, but 'sweet smells' like honeyed aromas in Rieslings, do not increase body…only the richness of aroma.]

Increasing sugar content almost always dictates an ever heavier and oilier wine body. This should have been evident in the shift from semi-sweet to sweet white wine. To really go over the top sometime, compare body between water and any desert wine like Sauternes! Whew! Its got the body of motor oil…all because of the sugar content!

For wine balance, this sugar factor works against the sharp acids and harsh tannins to soften the wine and pull it back towards the sweet side. See how hard this stuff is? It's like rocket science! But now it actually does get a little tricky…

Alcohol We all know what that is! But do you know how it affects the wine's taste and body? I know it seems counter-intuitive, but straight alcohol actually gives a wine a sweet-ish sensation! Even though pure alcohol is supposed to be odorless, colorless and tasteless, if you dilute a shot or two of high-proof alcohol with 6 ounces of water, you will detect a creamy, sweet sensation that is obviously not coming from the water.

In wine, increasing alcohol builds up more body. No surprises there. However with wine balance, alcohol counter-balances the acids and tannins slightly to pull the balance back toward the sweet side. This effect is most evident in red wines, as they have no sugars in them as we pointed out above. Without the alcohol to balance with its sweet sensation, all red wines would seem overly tannic and acidic and harsh.

Finally, very high levels of alcohol (as you probably have experienced doing straight shots of liquor) have a very 'hot' quality to them as well. When not adequately offset by tannins or acid, high alcohol makes the wine unbalanced and hot. It's also why red wines can be so much more alcoholic than whites, but can still stay in balance as the tannins smooth out that perception.   Alcohol is tricky on the palate to be sure.

So how does a wine achieve balance?

Now it gets easy: acids and tannins are on one side of the wine equation, with sugars and alcohol on the other. These forces work together… or against each other as the case may be…to create a balance. A wine 'feels right' and is in balance when the combination of these factors works together well. And every wine style is unique, so there is no single balanced equation to be achieved; meaning, some wines may have high proportions of acids and sugars which balance (many German Rieslings), while other wines may use high acids and high tannins to balance high sugar (Port), or even low levels of tannins to make a balance with low levels of alcohol (like a French rosé from Tavel).

For those of you that are thinking: "But that Port wasn't balanced, it seemed really sweet to me,", I can assure you that without the high levels of tannin and acid, that Port would seem a million times sweeter…to the point of being disgusting! Which brings up another point about balance…

If one of these factors is way over the top, or another is severely lacking, then things can seem terribly wrong. A total lack of acids make a wine seem 'flabby'; high levels of tannins without a balance can make a wine seem 'hot;' high levels of sugars without a balance make a wine seem 'sickeningly syrup sweet.'

As a quick made-up example: Wine #1 has 10 milligrams of residual sugar in it but no acids at all. Wine #2 has 20 milligrams of residual sugar, but also some acid character. Wine #1 will taste significantly sweeter, even though it has only half the sugar as wine #2. It's all about the counter-balancing effect of these ingredients, and the balance they create. You dig?

Now balance is dependent on the grapes for that year, the style of the wine, and the skill of each winemaker, so I cannot tell you if you drank a balanced wine in this exercise. In fact if you think about it, each of us has individual opinions about what we think tastes good. So to a certain degree, another factor of balance is the wine drinker's personal preference. I can't judge that for you, but we can get back to our primary lesson on body…

And what does any of that stuff have to do with body?

Okay you got all that balance stuff down? Then how do we understand those same factors in terms of their influence on body?

Simply put: the more stuff you add, the more body you get. I know. It's so simple as to be self-evident. Well, mostly… Increasing sugars, increasing acids, increasing alcohol, or increasing tannins in a wine all do result in an increased 'mouth feel', an increased 'heaviness' or 'chewiness', which I equate to increased body. (Some call it increased concentration, some call it increased complexity.) Without any of these factors, you pretty much just have water. And let me stress this: I don't mean you have the taste of water, I mean you have the feel of water. Thin, light, and, ummm… watery. Water has no body.

Even the lightest wines have *some* body. For instance, reconsider that Vinho Verde again. Hell, I've had some of those wines that were already so close to water that if I held my nose, I might not be able to tell the difference. But the addition just of those acids does add some body. The bite is there for us to distinguish in our mouth. It feels like something…. and that something gives it some body. And for that particular wine, it is almost solely just the acids.

For the semi-sweet versus the full-on sweet wine, you should have detected the sweeter wine feeling heavier in your mouth. Just adding that extra sugar makes a significant difference in how it feels…doesn't it? Again, I'm not talking about it tasting sweeter; I'm talking about how it *feels* heavier. If you still aren't getting it, then go back and try the Liebfraumilch and the Riesling Auslese again, but this time pinch your nose when you are trying it. Don't taste…*feel*. That Auslese is heavier, chewier, thicker…it has more body. Hell, it probably clings to the inside if the glass more too! That is some sticky body!

For the tannin example, can you now easily detect the heavier mouth feel on the young Rhône red versus the Tavel rosé? Granted, reality is never as simplistic as this. Depending on the bottles you have, the producers, and the vintage, I'm sure that there are other factors at play that are creating different bodies of those two wines. But you should definitely have been able to *feel* how that extremely tannic Rhône was way heavier on your palate than the pink wine. Did you get it?

And the alcohol factor? That's why I threw in the ringer Port. While the color, and therefore tannic component, of the Port looks similar to the Shiraz or Cab, the alcohol levels make all the difference. In fact, Port has high levels of sugar, tannins, and alcohol, so it is over the top for body and mouth feel! And as I suggested earlier, even though levels of all three are really high, in combination they tend to balance each other out. They are supposed to come into balance over time, which brings us to the last point…

Why aren't all wines just immediately crafted in perfect balance with perfect body right this second?

There are reasons why a winemaker would choose to intentionally beef up the sugars, acids, tannins or the alcohol of a wine, even when doing so makes for a wine of heavy body that is out of balance. Why?

Primarily: for longevity. Increasing any one of these factors makes a wine last longer, and gives it the ability to age. And I'm speaking mostly of red wines here, since most whites are made for immediate consumption. But even those have exceptions: some well-crafted German Rieslings have high acids and sugars which allow them to mature for decades, since the high concentrations of sugar and/or acid actually discourage bacterial growth. High alcohol levels have the same effect.

But it is mostly reds that go through this process for the one factor they have that whites lack: tannins. Many young red wines are aggressively tannic—so tannic you don't even want to drink them without food because they are just too much! "It's too young!" you will hear people say about a tannic wine. Since tannins in wine dissipate over time, a wine's balance and body change with age. And it is the aging process which gives wine its magic abilities. Only through slow aging can wines garner those absolutely fantastic qualities that make for legendary wines…and legendary prices!

So to make your wines last through the aging, a winemaker must craft a combination of acids and tannins and alcohol that he feels will fold together perfectly over time. The alcohol and sugars pretty much stay the same, but over time the tannins and acids soften considerably. That's why the crafting of age-worthy wines is such an art form! They are basically painting a canvas that doesn't show up for 10 years! Try that, Picasso!

Back to point: now you know why reds are usually aged while whites are consumed young… and why aggressively tannic wines simply must be aged. You also should understand why high acid white wines might last longer than low acid ones. And now you know the secret to why high sugar wines like German Beerenauslese and Icewein can last for a very long time. But which wine should you probably not even touch for twenty years after its creation? If you guessed high quality Ports, you are dead on! Why? Because they have high levels of *everything* which need time to mellow out and come into **balance**—and are what give them such heavy **body**.

In summary:

❖ Acid, tannins, sugars, and alcohol work in combination with each other—or against each other—to achieve **balance** in a wine.

❖ There is no perfectly balanced equation for all wines. Individual styles have different combinations of these base components, which is why there is such variation in the world of wine...even within the same types of wine from the same producer. Some may be heavier on the acid, while others heavier on the sugar.

❖ We mostly only fault a wine's balance when one of these components is entirely absent or another is way over the top.

❖ Increasing levels of any one of those components, or a combination of several components, also increases the **body**, or mouth feel of a wine.

❖ A winemaker may intentionally craft a wine that has very high levels of one or more of these components, in an effort to make a wine that will age and get better later as the components soften and come into balance.

❖ As a red wine ages, tannins and acids 'soften' in a process which alters both the body and balance of the wine.

What process or combination of processes the winemaker decides to use contributes to what flavors and aromas are in the finished wine, as well as its body and balance. Remember, most wine books and wine people would give you a whole separate lesson on heavy versus light body, a whole separate lesson on flavor balance, and a whole separate lesson on acid versus sugars, with a diagram of the human tongue thrown in for kicks. We've just done it all in one fail swoop, so now you can get back to drinking. It doesn't really matter if you memorize all the ins and outs of this balanced equation approach to body. I just want you to understand how it works so that you will be a smarter drinker, especially around other wine drinkers. At least you now know what they are referring to when someone says the wine seems 'thin' or 'full-bodied' or 'out of balance' or 'too hot'. Well...do you?

(Oh, for you non-drinkers, you can mimic this lesson with water, lemonade, cola, exceptionally dark green tea, grapefruit juice, tomato juice, and a shot of straight maple syrup. Have fun with that... losers! This book is for drinkers!)

	Crisp vs Creamy	Semi-sweet vs Really sweet	Rose vs Young, dark Rhone	Big round red vs Port
Body:				
Other comments:				

3.

WHITES

Did you make it through the Winemaker Influence section? Congrats! Now the party can really get started! For this section deals with the light and bright whites of the wine world, and your assignments will become increasingly easier for the rest of this drinking manual. You can do these lessons on any order that you wish, according to whatever variety strikes your fancy for that day. There won't be too many other big words or concepts to consider here, but what you do have to really focus on is the aromas and flavors. Work hard to differentiate the unique characteristics of these varietal white wines, and build your knowledge base. Like any artist, the more colors you have on your palate, the richer the picture you can paint. So fill your palate up with these...

Lesson 8: The Big Three Whites

Chardonnay, Sauvignon Blanc, & Riesling. What they smell like. What they taste like. Where you get them from. What foods bring out the best in them.

Are these the best three white grape varieties on the planet? Not necessarily, but they are certainly the three white grapes that are used to produce the vast majority of white table wines…the Zraly text says "more than 90 percent of all quality white wine is made from these three grapes," and a quick browsing of any wine store shelf would confirm this fact. They also happen to be the 'establishment' grapes for a host of classic wine regions, which is perhaps why they are still so prevalent in today's wine world: Burgundy and Chablis are whites made from Chardonnay; Sauvignon Blanc is a Bordeaux and Loire staple; and Alsace and most German exported wines are forever Riesling. So what's all the fuss with these three? Let's take a look.

What to grab:
- One bottle from each list of the following from your local wine shop or grocer:
  Riesling
    Bottles $5-$15
    - Richter Piesporter (Germany)
    - Yalumba Y Series (Australia)
    - St.-Urbans-Hof (Germany)
    - Hogue Columbia Valley (Washington)
    - Kukl Kabinett (Germany)
    - Mosselland Piesporter Michelsebrg Spatlese (Germany)
    Bottles $16-$30
    - Pacific Rim (Washington)
    - Hugel (France)
    Bottles $30+
    - Richter Graacher Himmelreich Auslese (Germany)
    - J.j. Prum Auslese Wehlener Sonnenuhr Riesling (Germany)
  Sauvignon Blanc
    Bottles $5-$15
    - Bogle (California)
    - Babich (New Zealand)
    - Brancott (New Zealand)
    - Quintay (Chile)
    - Simi (California)
    - Robert Mondavi Fume Blanc (California)
    - Kim Crawford (Australia)
    - Di Stefano (Washington)
    Bottles $16-$30
    - Cloudy Bay (New Zealand)
    - Domaine Cailbourdin Pouilly-Fume "Les Cris" (France)

> Domaine La Croix St. Laurent Sancerre (France)

Bottles $30+
> Alphonse Mellot Edmond Sancerre (France)

<u>**Chardonnay**</u>

Bottles $5-$15
> Butterfield Station (California)
> Post Road (Washington)
> Alamos (Argentina)
> Casa Lapostolle Casablanca Valley (Chile)
> Stephen Vincent (California)
> Greg Norman Estates Yarra Valley (Australia)

Bottles $16-$30
> William Fevre Chablis Champs Royaux (France)
> Chateau Ste. Michelle Cold Creek (Washington)
> Bernard Defaix Chablis (France)
> L'Ecole N⁰ 41 Columbia Valley (Washington)
> Louis Jadot Pouilly Fuisse (France)

Bottles $30+
> Frogs Leap (California)
> Cakebread (California)
> La Soufrandiere Pouilly Vinzelles Quarts (France)

Don't worry if you can't find these exact labels. You can always ask your local wine store salesperson to help you pick out a good representation of the varieties I have described above. Show them the list and tell them to get you something close to it.

What to do:

1. First, before you even chill them, open all three bottles immediately and pour a glass of each. It might be wise to label each glass accordingly (which varietal it is). Take the rest of both bottles and stick them in the fridge for 30-45 minutes.

2. While you are waiting for the wines to chill, go ahead and assess each of the three white wines, and it really makes no difference what order you do them in (*unless you purchased a wine that you know to be sweeter, in which case do that one last*.) What is the color, the aroma and the taste of each wine? Record all impressions. You should be able to detect a distinct difference in flavors and aromas in these wines…and perhaps even different colors as well, but that is not of great significance for our exercise. Really work hard at picking out and describing differences between each of these varietal wines. As always, include as many descriptors as possible to define each wine, even if some may be a stretch. Check below for a bit of varietal vocabulary to choose from.

3. Now pull out the chilled wines from the fridge and repeat Step 2. As you go back and forth from one wine to the next a few times, be sure to cleanse your palate with a bit of bread or plain crackers. Sip some water as well, if you absolutely must. But don't do any other food—it will easily alter the results of your descriptions. Don't

worry: you can finish the bottles with dinner later.

4. If you are feeling cocky, go ahead and compare the warm versus cool glasses for each varietal wine. Are you starting to pick up tangible differences in the wine smells and flavors as the temperatures change? If so, you rock. If not, you are still cool: just hang in there. You are drinking wine after all, and life is good!

5. Keep the rest of the bottles for dinner. Try with a variety of different foods. What works? What doesn't? What foods bring out more flavors in the wine? What foods overwhelm the wine? I want you to experiment as much as you want to see what works best for you, but listed on the next several pages are some variety/food pairings that do seem to bring out the best in these particular whites that you might want to try.

What to look for:

The whole point of this exercise, and others to follow, is to get you fluid and fluent in recognizing and describing the major varietal wines produced in the world today. Of course you know that the term *varietal wine* means a wine made predominately, if not exclusively, from one type of grape. As you do the tasting, focus on comparing and contrasting these different wines, with more of an emphasis on contrasting.

Well, congratulations! You have now officially tasted the world's work-horse white wines. They are never hard to find, and they come in a huge variety of styles, so you can drink a new one every day for the rest of your life and still never run out of new tastes. Some are bone dry, some are sweet. Some are made in stainless steel, some are heavily oaked. Some are fruity and delicate, some are full-bodied, acidic, or aggressive. Flavors and aromas are influenced by the climate, the particular growing season of the year of production, and winemaker style. And a lot of these differences are encapsulated by, or can be defined by, what region they are from. A California Chardonnay is a uniquely different beast than a Burgundy Chardonnay, but I'm hoping mostly that this exercise at least helped you differentiate between the Chardonnay, the Sauvignon Blanc, and the Riesling.

Later exercises in this Boyer tasting round-up will compare some varietal wines from different regions, but you should do them on your own as often as possible too. When purchasing a bottle of one of these whites for dinner, go ahead and grab an inexpensive version of the same grape from a different part of the world. Your knowledge will grow exponentially, even if it is at the expense of your wallet decreasing mathematically. But I digress. Let's look at some specifics…

The Specifics:

Let's take a look at each of these major varieties one at a time to provide some tasting and smelling clues, some regional background, and also some food paring hints. Let's start with the big boy…

Chardonnay

The most recognized white wine produced, hands down. Is typically light and accessible and has no dominant fruit characteristics that overpower the palate. Perhaps that's why it's so widely consumed: it's easy. This grape adapts well to a variety of climates, and this combined with market demand means that there is a whole ocean of Chardonnays to choose from out there, many of them mundane and mediocre. I often tell folks that you can always pick out a Chardonnay from just about any other varietal white wine in a blind tasting, because it's the one that doesn't taste like anything. However, winemaker influence plays a huge role in how the finished wine turns out, and Chardonnays can be acidic and minerally from Chablis, or tropical fruity and light from Australia, or oaky-buttery bombs from California. Hopefully, no matter which one you tried for this exercise, you were able to pick out some of these fruit descriptors:

Distinct Chardonnay descriptors: green apple, buttery, citrus, melon, oak, pineapple, toast, vanilla

General Chardonnay descriptors: apple, apricot, banana, cucumber, fruit salad, gooseberry, grapefruit, lemon, lime, mandarin, mango, melon, nectarine, peach, pineapple, rockmelon, honeydew melon, spicy, straw, tropical fruit.

Where Chardonnay is grown: Damn near everywhere anymore, but the principle regions that produce wines of distinction include Burgundy (specifically identify them with Chablis and Cote de Beaune), Champagne, California, Oregon, and Australia.

What foods are often paired with Chardonnay: Heavier oaked or very fruity Chards go well with mild to medium flavored cheeses, pork, crab, lobster, clams, mussels, seafood in light/white sauces, chicken, white fish, game bird, and pretty much any pasta dish with a light cream sauce. More acidic cool climate Chards, like from France, go well with mild cheeses, light appetizers, oysters, shrimp, clams, mussels, seafood in light sauces, salmon, tuna, pasta dishes with heavier white sauces, pork, veal, and chicken.

And let's do some specific cheeses too
Definitely do: Triple Cream Brie, Brunet, Tete De Moine
Will work: Caruchon, Epoisses, Humboldt Fog
Possibly try: Catija, Bel Paese, Mild Cheddar, Panela, Brick

Sauvignon Blanc

A much lesser known variety than its big brother Chardonnay, Sauvignon Blanc is still a principle grape of many classic European wines, and has increasingly made its way into the New World line-up as well: Just check out New Zealand, which is gaining a reputation for great Sauvignon Blancs. Generally fresh, clean and fruity wines, usually with some serious acidity, these guys express much more of their fruit character and succumb less to intensive winemaker manipulations. That is, you don't typically find them 'buttered up' or heavily oaked. And I must warn you: Sauvignon Blancs have a reputation of being very grassy, very herbaceous (tastes like you are chewing on leaves) and the ever popular descriptor of cat urine is easily discovered if you end up tasting more than a few of these in your lifetime. This is one of the few whites that I personally think almost demands food. I say grill anything, and give it a go.

Distinct Sauvignon Blanc descriptors: grapefruit, grass, gooseberry, lemon, herbs, cat pee (no, this is not a misspelling), green olive

General Sauvignon Blanc descriptors: apricot, blackcurrant bud, freshly cut grass, fresh peas, gooseberry, grapefruit, grassy, green apple, green bean, green pepper, green plum, gunflint, herbaceous, lantana, lemon, lime, melon

Where Sauvignon Blanc is grown: Bordeaux and Loire in France, Northeastern Italy, Marlborough in New Zealand, Washington, California, South Africa, Chile.

What foods are often paired with Sauvignon Blanc: mild cheeses, goat cheeses, grilled pork, scallops, crab, lobster, clams, mussels, pretty much all shellfish, seafood in light/white sauces, smoked salmon, grilled sea bass, grilled shrimp, grilled chicken, grilled tuna, chicken or veal in cream sauces, all white fish, light cheese soufflés, and pretty much any pasta dish with a light cream sauce. Do you like it hot? Sauvignon Blanc is a lot of folks' first line of cooling defense from spicy hot Thai, Vietnamese, and Indian dishes too.

And let's do some specific cheeses too
Definitely do: Crottin, Le Chevrot
Will work: Chabichou Du Poitou, Bucheron
Possibly try: Valencay, Double Gloucester, Goat Cheese, Teleme, Neufchatel

Riesling
The uber-grape of Germany, Rieslings don't have quite the following in today's wine world as they used to enjoy, but they are on the rebound. Perhaps it is the honeyed, floral scent of the grape that consumers immediately associate with a sweet wine that has put consumers off, as such wines are not in vogue right now. However, you now know that this is not always the case: Rieslings are produced across the entire range of sweetness levels, and coupled with their savory aromas make an easily recognized and delightful beverage for sipping or food pairing. You should have no trouble at all differentiating these wines from other varietals after you complete this lesson.

Distinct Riesling descriptors: fruity, honeyed, honeysuckle, floral
General Riesling descriptors: apricot, blossom, citrus, honey, honeydew,

honeysuckle, lemon, lime, peach, tropical fruit, rose petal, flower, floral, berry, diesel, lanolin, petrol, kerosene, gunmetal, geranium, rose petal, orange blossom, jasmine, tropical fruit.

Where Riesling is grown: Like Chardonnay, this grape/wine can be found far and wide, just not in as great a number. Classic areas include Mosel-Saar-Ruwer, Rheingau, Rheinhessen and Pfalz in Germany; Alsace in France; New York, California, Oregon, Washington and South Australia. And watch out for great deals on Rieslings coming out of Austria too.

What foods are often paired with Riesling: And here I'm speaking primarily of dry to semi-sweet versions of Rieslings: mild cheeses, pates, light creamy dips, seafood in light sauces, salmon, light tuna, chicken or veal in cream sauces, duck, goose, freshwater fishes like trout, salads with a light vinaigrette dressing, and definitely try cream soups too. For sweeter version of Riesling: mild cheeses, mild curries, any light meat dish flavored with sesame or ginger (as long as its not too spicy), roast venison, smoked salmon, more cream soups, and any white meat dish stuffed with or accompanied by fruit—caramelized or cooked fruit on the dish is the bomb. And you may have not had it yet in your life, but you will: foie gras with a sweet Riesling is to die for. Don't know what that is? Look it up!

And let's do some specific cheeses too
Definitely do: Alpkase, Adrahan Farmhouse, Morbier
Will work: Gruyere, Chesire, Colby, Monterey Jack
Possibly try: Bandaged Cheddar, Clothbound Cheddar, Rogue River Blue

<u>**Additional barrel flavors**</u> you may detect in any of the wines depending upon how much time it spent in oak: (From Lesson 4: Got Wood? The Influence of Oak)
  ➤ *Woody descriptors:* Cedar, charred wood, cigar box, coffee, green wood, oaky, pencil shavings, resinous, sandalwood, sawdust, smoky, toasty, tree bark, vanilla.
  ➤ *Barrel flavor (malolactic fermentation) descriptors:* Almond, banana, burnt caramel, butter, buttered popcorn, butterscotch, cashew, coconut, lanolin, vegemite
  ➤ *Lees flavor descriptors:* Baked bread, creamy, cheesy, leesy, nutty, yeasty, yoghurt

One final note: Please keep in mind the tremendous variation that exists among each type of varietal wine. As stressed several times now, an Australia Chardonnay can be radically different than a French Chardonnay. And a very sweet Riesling is radically different from a bone dry Riesling, even if they are from the same region. That's why I decided to use the grape varieties as a backbone for your wine knowledge, that you can then add appendages to later. Try to focus on the fruit components as you do this exercise, knowing full well that everything else changes according to region, climate, and winemaker influence. Good luck. Many more varieties to come…

	Comments
Chardonnay	
Color:	
Bouquet (smell):	
Taste:	
Body:	
Compare/Contrast to other wines:	
Sauvignon Blanc	
Color:	
Bouquet (smell):	
Taste:	
Body:	
Compare/Contrast to other wines:	
Riesling	
Color	
Bouquet (smell):	
Taste:	
Body:	
Compare/Contrast to other wines:	

Lesson 9: The Lighter Whites

Chenin Blanc, Pinot Grigio, & Semillon. What they smell like. What they taste like. Where you get them from. What foods bring out the best in them.

After the three big whites of the previous lesson, these grapes/wines are definitely a tier down the ladder both in terms of quantities produced and name recognition. However like everything in the wine world, that may be changing fast; Pinot Grigio in particular seems to be on the war path in terms of gaining consumer confidence, and all three of these grapes are important regional players in a host of European wine districts. Be forewarned: there are dozens of other white varieties that are gaining ground too, but I still think you need to know these three first in order to fill out your knowledge palate of major white wines. And these wines are a bit trickier too: unlike your previous exercise, Chenin Blanc, Pinot Grigio, and Semillon are more subtle and don't have any radical aromas or flavors that jump out and grab you. You'll have to work a bit harder to differentiate these from each other, but I have confidence that you can do it! As you go through this exercise, try to remember the big three whites you did last time, and compare and contrast your notes as well.

What to grab:
* One bottle from each list of the following from your local wine shop or grocer:
  <u>Chenin Blanc</u>
    Bottles $5-$15
    * Ken Forrester Stellenbosch (South Africa)
    * Barton & Guestier Vouvray (France)
    * Man Vintners (South Africa)
    * Pine Ridge (California)
    * Domaine Pichot Vouvray (France)
    * L'Ecole Nº 41 Walla Voila (Washington)
    Bottles $16-$30
    * Champalou Vouvray Sec Tendre (France)
    * Domaine Francois Pinon Vouvray (France)
    Bottles $30+
    * Domaine Huet Vouvray Le Mont Demi-Sec (France)
    * Nicolas Joly/Coulee de Serrant Savennieres Coulee de Serrant 1995 (France)
  <u>Pinot Grigio</u>
    Bottles $5-$15
    * Beringer (California)
    * Busa De Lele (Italy)
    * Foris Rogue Valley (Oregon)
    * Ermacora (Italy)
    * Rene Mure 'Tradition' (France)
    * Tommasi (Italy)

Bottles $16-$30
- ➤ Jermann Pinot Grigio (Italy
- ➤ Santa Margherita (Italy)
- ➤ Trimbach Pinot Gris Reserve (France)

Bottles $30+
- ➤ Trimbach Pinot Gris Hommage a Jeanne (France)
- ➤ Robert Mondavi, 'La Famiglia' 1998 (California)

Semillon

Bottles $5-$15
- ➤ Cricket Pitch White (Semillon/Sauv Blanc blend, classic Bordeaux-style white) Australia
- ➤ Chateau Ducasse White Bordeaux (France)
- ➤ Chateau La Colline (France)

Bottles $16-$30
- ➤ L'Ecole Nº 41 Semillon (Washington)
- ➤ Brokenwood (Australia)
- ➤ Torbreck Woodcutter's (Australia)

Bottles $30+
- ➤ Morlet La Proportion Doree (Semillon/Sauv Blanc blend, classic Bordeaux-style white) (France)

Don't worry if you can't find these exact labels. You can always ask your local wine store salesperson to help you pick out a good representation of the varieties I have described above. Show them the list and tell them to get you something close to it.

What to do:

1. First, before you even chill them, open all three bottles immediately and pour a glass of each. It might be wise to label each glass accordingly (which varietal it is). Take the rest of both bottles and stick them in the fridge for 30-45 minutes.

2. While you are waiting for the wines to chill, go ahead and assess each of the three white wines, and it really makes no difference what order you do them in (*unless you purchased a wine that you know to be sweeter, in which case do that one last*.) What is the color, the aroma and the taste of each wine? Record all impressions. You should be able to detect a distinct difference in flavors and aromas in these wines…and perhaps even different colors as well, but that is not of great significance for our exercise. Really work hard at picking out and describing differences between each of these varietal wines. As always, include as many descriptors as possible to define each wine, even if some may be a stretch. Check below for a bit of varietal vocabulary to choose from.

3. Now pull out the chilled wines from the fridge and repeat Step 2. As you go back and forth from one wine to the next a few times, be sure to cleanse your palate with a bit of bread or plain crackers. Sip some water as well, if you absolutely must. But don't do any other food—it will easily alter the results of your descriptions. Don't worry: you can finish the bottles with dinner later.

4. If you are feeling cocky, go ahead and compare the warm versus

cool glasses for each varietal wine. Are you starting to pick up tangible differences in the wine smells and flavors as the temperatures change? If so, you rock. If not, you are still cool: just hang in there. You are drinking wine after all, and life is good!

5. Keep the rest of the bottles for dinner. Try with a variety of different foods. What works? What doesn't? What foods bring out more flavors in the wine? What foods overwhelm the wine? I want you to experiment as much as you want to see what works best for you, but listed on the next several pages are some variety/food pairings that do seem to bring out the best in these particular whites that you might want to try.

What to look for:

The whole point of this exercise, and others to follow, is to get you fluid and fluent in recognizing and describing the major varietal wines produced in the world today. Of course you know that the term *varietal wine* means a wine made predominately, if not exclusively, from one type of grape. As you do the tasting, focus on comparing and contrasting these different wines, with more of an emphasis on contrasting. After you finish…

You're done! How was it? Could you tell them apart? While these three whites don't have the flexibility and massive range of styles that the 'Big Three' whites do, you can still find some major stylistic differences in how these wines are made in different parts of the world. Semillon in particular can be produced fruity and dry (as they do it in Hunter Valley, Australia) or it can be crafted into a complex, lemony affair (like in some of the greatest whites of Bordeaux), or even into a full-fledged sweet-ass wine (as they do in Sauternes.) Chenin Blanc is a chameleon that takes on an array of fruity characteristics and sweetness depending upon where its grown. And Pinot Grigio? Don't get me started! While its claim to fame may be as an insipidly easily sipping wine, it transforms itself at the drop of a hat in terms of its subtle flavors and name changes. Every region that grows the stuff calls it something different, but Pinot Grigio may be the name that wins out here in the 21st century. But I digress as usual…

Later exercises in this Boyer tasting round-up will compare some varietal wines from different regions, but you should do them on your own as often as possible too. When purchasing a bottle of one of these whites for dinner, go ahead and grab an inexpensive version of the same grape from a different part of the world. Your knowledge will grow exponentially, even if it is at the expense of your wallet decreasing mathematically.   Let's look at some specifics…

The Specifics:

Let's take a look at each of these major varieties one at a time to provide some tasting and smelling clues, some regional background, and also some food paring hints. Let's start with…

Chenin Blanc

You probably haven't heard of this one...but you're well on the way to true connoisseurship after having tried it. It's a staple of the Loire region, and anytime you see a wine labeled Vouvray (the region), that's 100% Chenin. Its light and fruity with a implied hint of sweetness and honey on the nose. In cool climates like France it maintains an acid edge that balances the wine nicely when it's made into semi-sweet or sweet versions. As with most whites, warmer climates bring out more of the fruit and less of the acid crispness.

Distinct Chenin Blanc descriptors: apples, apricots, pear, nuts, honey, marzipan

General Chenin Blanc descriptors: apple, floral, grassy (light), herbaceous, herbal, tropical fruit, almond, honey, lanolin, pear.

Where Chenin Blanc is grown: In France, anything from a place called Vouvray is 100% Chenin Blanc. Vouvray is a region in the Loire Valley. In the US, the Central Valley of California grows some fruity expressions of Chenin and the cooler Washington state climate has ones closer to the French style. Some good examples are coming out of South Africa, where they call the wine/grape Steen.

What foods are often paired with Chenin Blanc: Mild cheeses and all kinds of sliced fruit (stuff like honeydew and cantaloupe, not so much citrus), steamed shrimp, grilled shrimp, shrimp scampi, lobster, clams, mussels, seafood in light/white sauces, game bird, pork in light sauce, veal, baked ham, grilled ham. Damn, ham is good. Fruitier versions from California can also be paired with light appetizers, pates, dips, oysters, duck, semi-spicey Asian dishes, and light pastas. Vouvrays are specifically awesomely matched with shellfish in northern France.....mmmm...mussels and Vouvray. The Frenchies got that right for sure!

And let's do some specific cheeses too
Definitely do: Banon, Blue Castello
Will work: Derby, Graddost, Pecorino
Possibly try: Provolone, Mozzarella, Swiss, Colby

Pinot Grigio

Where did this guy come from? Seems like this grape is one of the hot new commodities in the wine world, and mark my words: come this summer, you'll not go to a single dinner party where its not being served well-chilled on the veranda. But it's not that new, and it's not just called Grigio. Alsace, France has been producing the stuff for years under the name Pinot Gris, and their wine is a medium to full-bodied affair that can have some assertive flavor. Especially compared to the Italian and California versions which are a bit lighter, refreshing, and not-so-complex, which of course makes them comfortably quaffable.

Distinct Pinot Grigio descriptors: pear, spring blossom, nougat, light spice, wet straw, flinty, even hints of granite!

General Pinot Grigio descriptors: apple, apricot, citrus, forest floor, hay, honeysuckle, minerally, musk, nuts, nougat, pear, perfumed, pineapple, roses, violets, wild honey, wet straw, savory.

Where Pinot Grigio is grown: We now think of this grape as an Italian affair, and it's true that its been historically produced in northeast Italy, and more recently in Tuscany. But it's also in Alsace in France, Baden in Germany, and Oregon and California in the US.

What foods are often paired with Pinot Grigio: mild cheeses, grilled pork, seafood in light/white sauces, seafood in heavier sauces, chicken, light pasta dishes, lightly spiced Asian cuisine, ham, and really any other cold cuts, cut fruit and light picnic fare that you would throw together on a sunny afternoon to take up to the parkway. Party.

And let's do some specific cheeses too
Definitely do: Trugole, Robiola due Latti , Humboldt Fog
Will work: Stracchinos, Mt Tam, Aged Provolone, Echo Mountain
Possibly try: Mahon, Colby, Idiazabel, Swiss, Provolone

Semillon
Probably never heard of this one either, right? That's cool. Hopefully you had a good basic one, either dry or with a bit of sweetness. But more than that, I hope you someday break down and purchase a Sauternes, which is a unique dessert wine from Bordeaux made from 100% Semillon. And speaking of Bordeaux, Semillon is blended with Sauvignon Blanc to create the classic whites of Graves. Australia and California produce some dry, fruity, medium-bodied ones which can have intense hints of peach or apricot. As the sweetness levels increase in this varietal wine, so do the viscosity and flavor dimension. Give it a go.

Distinct Semillon descriptors: canned sweet corn, peaches, ripe apricot, grassy, lemon, lanolin, honey, toast

General Semillon descriptors: apple, apricot, citrus, flinty, gooseberry, grassy, green beans, herbaceous, lantana, lemon, lime, olive, passion fruit, peach, pea pod, pear, pineapple, quince, raisin, saffron, spicy, tropical fruit, asparagus, fig, honey.

Where Semillon is grown: Classic white grape of Bordeaux, France, specifically Graves, Barsac, and Sauternes. Hunter Valley in Australia. California and Washington state in the US.

What foods are often paired with Semillon: mild cheeses, goat cheeses, grilled pork, scallops, crab, lobster, clams, mussels, pretty much all shellfish, grilled sea bass, shad roe, goat cheese soufflé, oysters on the half-shell, fresh crab salad, mussels, clams, chilled lobster consommé, scallops, grilled shrimps, roasted chicken, grilled chicken, game birds, duck, goose, heartier pasta dishes, grilled pork, veal, veal Parmesan, and some more picnic cold cuts too.

With the ultra-sweet dessert versions like Sauternes: any desert dish that is just slightly sweeter than the wine, mild cheeses, any white meat dish stuffed with or accompanied by fruit—caramelized or cooked fruit on the dish in particular, And you may have not had it yet in your life, but you will: foie gras with a sweet Sauternes is to die for. Don't know what that is? Look it up!

And let's do some specific cheeses too
Definitely do: Laura Chanel, Kasseri, Caerphilly
Will work: Camembert, Ubriaco, Couronne Lochoise
Possibly try: Brie , Mozzarella, Manouri, Feta

<u>**Additional barrel flavors**</u> you may detect in any of the wines depending upon how much time it spent in oak: (From Lesson 4: Got Wood? The Influence of Oak)
   ➢ Woody descriptors: Cedar, charred wood, cigar box, coffee, green wood, oaky, pencil shavings, resinous, sandalwood, sawdust, smoky, toasty, tree bark, vanilla.
   ➢ Barrel flavor (malolactic fermentation) descriptors: Almond, banana, burnt caramel, butter, buttered popcorn, butterscotch, cashew, coconut, lanolin, vegemite
   ➢ Lees flavor descriptors: Baked bread, creamy, cheesy, leesy, nutty, yeasty, yogurt

One final note:
Try to focus on the fruit components as you do this exercise, knowing full well that everything else changes according to region, climate, and winemaker influence. Good luck. Also try to compare back to your previous lesson and think about those other varietals wines you have already tried. What makes each one of them unique from each other? What foods bring out different flavors in each? Keep pushing yourself. Many more varieties to come…

	Comments
Chenin Blanc	
Color:	
Bouquet (smell):	
Taste:	
Body:	
Compare/Contrast to other wines:	
Pinot Grigio	
Color:	
Bouquet (smell):	
Taste:	
Body:	
Compare/Contrast to other wines:	
Semillon	
Color	
Bouquet (smell):	
Taste:	
Body:	
Compare/Contrast to other wines:	

Lesson 10: The Fantastic Floral Whites

Albariño, Gewürztraminer, & Viognier. What they smell like. What they taste like. Where you get them from. What foods bring out the best in them.

Now you've got 6 big white grapes/white wines under your belt, and hopefully you are fluent at not only telling them apart, but also at picking out some stylistic differences of them from region to region. And perhaps you've had some difficulties: perhaps some of those whites tasted or smelled similar, or had very similar characteristics. But now my friends you can relax with this lesson. Because these three whites are so fantastically floral that it will be impossible to mistake them for anything that has come before them. They are intense. They are spicy. They are a bouquet of flowers ready to explode into your nose!

BUT, they are also largely misunderstood. As I hope you remember from Lesson #5 (Sweet vs. Dry), just because a wine smells honeyed or sweet or spicy does not necessarily mean that it tastes that way. And most versions of Albariño, Gewürztraminer and Viognier are on the dry side, making them excellent pairs for a variety of foods and situations despite their powerhouses noses. In terms of complexity, raciness, and intrigue, these wines beat out the more mundane Chardonnays and Pinot Grigios every time for me. Know your floral whites, embrace the smells, and be bold enough to drink them with foods or on their own…and you will be florally fluent and fantastically fashionable.

What to grab:
- One bottle from each list of the following from your local wine shop or grocer:
  Albariño
  - **Bottles $5-$15**
    - ➢ Valminor (Spain)
    - ➢ Salneval (Spain)
    - ➢ Nessa (Spain)
    - ➢ Burgans (Spain)
  - **Bottles $16-$30**
    - ➢ Albarino D Fefinane (Spain)
    - ➢ Pazo Senorans (Spain) (my personal favorite on the planet)
    - ➢ Do Ferreiro (Spain)
    - ➢ Havens (California)
  - **Bottles $30+**
    - ➢ Lusco (Spain)

  Gewürztraminer
  - **Bottles $5-$15**
    - ➢ Foris (Oregon)
    - ➢ Pacific Rim (Oregon)
    - ➢ Hogue Columbia Valley (sweet) (Washington)
    - ➢ Fetzer (California)

> Valckenberg (Germany)

Bottles $16-$30
> Mader Alsace (France)
> Trimbach (France)
> Hugel (France)
> Adam, Jean-Baptiste Vin D'Alsace Reserve (France)

Bottles $30+
> Domaine Zind Humbrecht Heinbourg 1998 (France)
> J. Hofstatter Kolbenhof (Italy)

Viognier

Bottles $5-$15
> Smoking Loon (California)
> McManis (California)
> Yalumba Viognier The Y Series (Australia)
> The White Knight (California)

Bottles $16-$30
> Kunde Estate (California)
> Qupe (California)
> Arrowood Saralee's Vineyard (California)

Bottles $30+
> Guigal Condrieu (France)
> Alban Vineyards (California)

Don't worry if you can't find these exact labels. You can always ask your local wine store salesperson to help you pick out a good representation of the varieties I have described above. Show them the list and tell them to get you something close to it.

What to do:

1. First, before you even start to chill, open all three bottles immediately and pour a glass of each. It might be wise to label each glass accordingly (which varietal it is). Take the rest of both bottles and stick them in the fridge for 30-45 minutes.

2. While you are waiting for the wines to chill, go ahead and assess each of the three white wines, and it really makes no difference what order you do them in (*unless you purchased a wine that you know to be sweeter, in which case do that one last*.) What is the color, the aroma and the taste of each wine? Record all impressions. You should be able to detect a distinct difference in flavors and aromas in these wines…and perhaps even different colors as well, but that is not of great significance for our exercise. Really work hard at picking out and describing differences between each of these varietal wines. As always, include as many descriptors as possible to define each wine, even if some may be a stretch. Check below for a bit of varietal vocabulary to choose from.

3. Now pull out the chilled wines from the fridge and repeat Step 2. As you go back and forth from one wine to the next a few times, be sure to cleanse your palate with a bit of bread or plain crackers. Sip some water as well, if you absolutely must. But don't do any other food—it will easily alter the results of your descriptions. Don't worry: you can finish the bottles with dinner later.
4. If you are feeling cocky, go ahead and compare the warm versus cool glasses for each varietal wine. Are you starting to pick up tangible differences in the wine smells and flavors as the temperatures change? If so, you rock. If not, you are still cool: just hang in there. You are drinking wine after all, and life is good!
5. Keep the rest of the bottles for dinner. Try with a variety of different foods. What works? What doesn't? What foods bring out more flavors in the wine? What foods overwhelm the wine? I want you to experiment as much as you want to see what works best for you, but listed on the next several pages are some variety/food pairings that do seem to bring out the best in these particular whites that you might want to try.

What to look for:

The whole point of this exercise, and others to follow, is to get you fluid and fluent in recognizing and describing the major varietal wines produced in the world today. Of course you know that the term *varietal wine* means a wine made predominately, if not exclusively, from one type of grape. As you do the tasting, focus on comparing and contrasting these different wines, with more of an emphasis on contrasting. After you finish...

You're done! How was it? Could you tell them apart? It should have been no problemo.

While these three whites don't have the flexibility and massive range of styles that the 'Big Three' whites do, they are a whole lot of fun for various reasons. First, they are extremely hard to pronounce, which means you will be an automatic expert by simply being able to properly project their names into conversation. Second, they are not as widely produced or well known, but soon will be. Third, they are so pungent and floral that they scare off most folks who are unwilling to venture into un-chartered territory. And that is a shame, because these are superior summer sipping wines in my opinion, and they are great food matches as well.

Later exercises in this Boyer tasting round-up will compare some varietal wines from different regions, but you should do them on your own as often as possible too. When purchasing a bottle of one of these whites for dinner, go ahead and grab an inexpensive version of the same grape from a different part of the world. Your knowledge will grow exponentially, even if it is at the expense of your wallet decreasing mathematically. Let's look at some specifics...

The Specifics:

Let's take a look at each of these major varieties one at a time to provide some tasting and smelling clues, some regional background, and also some food paring hints. Let's start with...

Albariño

This is a wine grape on the rise....meaning you don't see hardly any versions of it from around the world right now, but I'm betting you soon will. Distinct, intense floral and fruit notes on the nose with a delightful roundness and body on the palate, these wines make Chardonnay seem like a wimpy little step-child. Albariño is the primary grape used to make dry white wine in the *Rias Baixes* (Lower Inlets) section of the *Galicia* region of Northwestern Spain. Considered by many to be Spain's premier quality white wine, Albariño is also known in Portugal as *Alvarinho* and often used as a component of Vinho Verde. They also typically have a serious acidic backbone which makes them great with seafood, but also makes them prone to premature dilapidation—meaning that you should drink them while they're new and young and fresh. Don't let these guys hang around the cellar too long...if at all!

Distinct Albariño descriptors: ripe apple, peach, citrus, flowers, almonds

General Albariño descriptors: apple, apricot, almond, almond paste, citrus, floral, grass (light), minerally, tropical fruit, refreshing, fresh, tart, lightly spritzy.

Where Albariño is grown: Almost exclusively an Iberian affair right now. We typically associate wines of Spain with hot, dry climates which are responsible for so much of their powerhouse red portfolio. But Rias Baixes is a region in northwestern part of the country, right on the coast, that has a much cooler and wetter climate...which is why this place is known for its whites with fruity floral noses. And the cool growing season is also why they can have almost biting acidity, much like their French white wine counterparts. A few are popping up in northern Portugal too, and it seems to me that those Portuguese expressions are even more acidic than the Spaniards. Try a Vinho Verde some time to see what I mean. Bracing!

What foods are often paired with Albariño: Mild cheeses, steamed shrimp, grilled shrimp, shrimp scampi, oysters, crab, lobster, clams, mussels, seafood in light/white sauces, game bird, chicken in light sauce, chicken in spicy sauces, hearty pasta dishes, baked ham, grilled ham. Go for the gusto some time and try it with spicy Asian dishes as well.

And let's do some specific cheeses too
Definitely do: Caprino Stagionato, Petilla, La Serena
Will work: Camembert, Comte, Caerphilly, Baked Brie

Possibly try: Colby, Jarlsburg, Pont' Leveque

Gewürztraminer
Dude! I love this stuff! What a kick ass nose…like burying your face in a bouquet of flowers. Gewürztraminer is one of the most pungent wine varietals, easy for even the beginning taster to recognize by its heady, aromatic scent. Fancy-smancy wine texts often report that "gewürz" translates from German as "spicy", but considering the list of various synonyms, the more likely contextual meaning is "perfumed". Either way, don't be put off by the nose; most of these wines are dry, even though they smell 'sweet,' especially those from Alsace (France). A bit of sweetness variation can be found though, as some semi-dry/semi-sweet versions are produced in Germany and across the New World.

Distinct Gewürztraminer descriptors: rose petal, gardenia, honeysuckle, lychee, linalool, peach, mango, spice, perfume. (Never heard of lychee? Pick up a can of lychee fruit at a grocery store that has a broad selection of international foods. This Asian fruit/nut is the most common descriptor of well-crafted Gewürztraminers.)

General Gewürztraminer descriptors: aromatic, cinnamon, clove, floral, lavender, lime, lychee, passion fruit, perfumed, potpourri, roses, spicy, honey, honeysuckle, minerally, violets, wild honey.

Where Gewürztraminer is grown: As previously alluded to, Alsace in northern France has some great dry expressions of these wines. Just over the border in Germany you will find a great variety of them too, sometimes labeled shorthand as Traminer. And you can find them in small supply from just about everywhere in the New World too: Oregon, Washington, and Australia all make noteworthy versions.

What foods are often paired with Gewürztraminer: These wines are an excellent match for fresh fruit and cheeses and a good complement to many simple fish and chicken dishes, especially recipes that include capsaicin (hot pepper) spices, oriental five spice, or even curry. In fact, they are probably your best bet for wine to pair with any kind of hot Asian cuisine: a rarity in the wine world altogether. Pates, dips, salmon, tuna, lightly peppered beef dishes, grilled pork, baked pork, grilled ham, baked ham…oh, and any distinctly German meat dish like sauerbraten go wonderfully with a nice Gewürzt.

And let's do some specific cheeses too
Definitely do: Appenzeller, Haystack Peak, Pecorino
Will work: Clothbound Cheddar, Petit Frere, St Pete's Select
Possibly try: Boursin, Chevre, Garrotxa, Swiss, Wensleydale

Viognier
Nobody except real French wine-ophiles even knew this grape ten years ago, but now it is storming into production across select parts of the

New World. The increasing acreage of Viognier in California, Virginia, Australia, and France is due partly to the consumers' growing interest in such exotic wines, and also because clonal selection has made this once-troublesome grape much easier to grow. A well-crafted Viognier can intoxicate one's senses the moment the bottle is opened. It offers up an intense exotic perfume of mayflowers and tropical fruits, and your palate can easily become overwhelmed by its richness of body and flavor. This richness of smell, like all the others this lesson, suggests a sweet wine, despite its usual complete dryness. But watch out! As these wines are still kind of 'new,' (in terms of the lack of experience that many winemakers and regions have with them) it is very easy to come across a boring or bland version of Viognier. If this happens to you, try another one from a different region or winemaker before you give up on them. Got a lot of bucks? Try the original 100% Viognier from a little region in the Rhone area of France named Condrieu. Not so loaded? Go for a Virginia or California label…many are surprisingly well-crafted..

Distinct Viognier descriptors: floral, orange blossom, acacia, violet, honey, apricot, mango, pineapple, guava, kiwi, tangerine, anise, mint, mown hay, tobacco

General Viognier descriptors: anise, apricot, blond tobacco, citrus, flinty, honey, jasmine, guava, kiwi, ripe pears, lemon, lime, musk, orange blossom, orange peel, violets, white peach, passion fruit, quince, spicy, tropical fruit.

Where Viognier is grown: Rare but classic white grape of Rhone Valley in France. Viognier is the single permitted grape variety in the famous appellations of Condrieu and Château Grillet, which are located on the west bank of the Rhône River, south of Lyon. The wines of Condrieu are the most famous, and most expensive, Viogniers in the world. Now, California and Australia are increasing their production of this wine, and lo and behold!: The great state of Virginia is turning up as a major producer of this varietal wine.

What foods are often paired with Viognier: mild to strong cheeses, goat cheeses, grilled pork, scallops, crab, lobster, clams, mussels, seafood in light sauces all the way to seafood in heavier sauces, (although I think I would keep shrimps out of the picture for this wine), fresh crab salad, roasted chicken, grilled chicken, game birds, wild game, anything gamey; game on, venison, lamb, duck, goose, heartier pasta dishes, veal, and picnic cold cuts too.

And let's do some specific cheeses too
Definitely do: Brillat-Savarin Petite, Natural Rind, Piave
Will work: Livarot, Berkswell, Lumiere
Possibly try: Hudson Valley Camembert, Pierre Robert

**Additional barrel flavors** you may detect in any of the wines depending upon how much time it spent in oak: (From Lesson 4: Got Wood? The Influence of Oak)

> ➤ Woody descriptors: Cedar, charred wood, cigar box, coffee, green wood, oaky, pencil shavings, resinous, sandalwood, sawdust, smoky, toasty, tree bark, vanilla.
> ➤ Barrel flavor (malolactic fermentation) descriptors: Almond, banana, burnt caramel, butter, buttered popcorn, butterscotch, cashew, coconut, lanolin, vegemite
> ➤ Lees flavor descriptors: Baked bread, creamy, cheesy, leesy, nutty, yeasty, yoghurt

One final note:
Try to focus on the fruit components as you do this exercise, knowing full well that everything else changes according to region, climate, and winemaker influence. Good luck. Also try to compare back to your previous lesson and think about those other varietals wines you have already tried. What makes each one of them unique from each other? What foods bring out different flavors in each? Keep pushing yourself. Many more varieties to come…

	Comments
Albariño	
Color:	
Bouquet (smell):	
Taste:	
Body:	
Compare/Contrast to other wines:	
Gewürztraminer	
Color:	
Bouquet (smell):	
Taste:	
Body:	
Compare/Contrast to other wines:	
Viognier	
Color	
Bouquet (smell):	
Taste:	
Body:	
Compare/Contrast to other wines:	

4.

REDS

You have arrived in heaven now... If you are not already a huge fan of red wines, you soon will be. Many wine drinkers feel that the first duty of any wine is to be red. And there does seem to be a natural wine drinking evolution from light whites to heavier whites to light reds to bold reds. Some reds have power and prestige, while others are more intricate and delicate, but all are worthy of our attention!

As with all chapters of this book, you can do these red wine lessons in any order that you want. However, due to increased alcoholic strength as well as the increased prices of these wines, I highly recommend group play with these big red varietal wines. And for more than just practical reasons: wine drinking is a social thing, and so you really should do it with some other folks. Drinking these wines as a group activity will not only help your pocketbook and possible hangover, but encourages dialogue and diatribes of waxing poetic about these lofty libations. Interaction with others is actually one of the most critical components of furthering your wine education and wine confidence, so grab some big reds and a group of friends and get to work...

Lesson 11: Bordeaux's Big Boy Reds: Cabernet Sauvignon, Merlot, & Cabernet Franc

Lesson 12: Burgundy's Big Hitter: Pinot Noir

Lesson 13: Italian Stallions: Sangiovese, Nebbiolo & Barbera

Lesson 14: Spanish Princes: Grenache & Tempranillo

Lesson 15: The Rhône Rangers: Syrah & Mourvèdre

Lesson 16: Americas' Finest: Zinfandel & Petite Syrah & Malbec

Lesson 11: Bordeaux's Big Boy Reds: Cabernet Sauvignon, Merlot, & Cabernet Franc

What they smell like. What they taste like. Where you get them from. What foods bring out the best in them. And what the hell is the Bordeaux Blend?

Well, here we go! Let's hit the reds! And we will bite off way more than we can chew with the biggest, baddest, hardest tasting of the whole damn book: Bordeaux. Why is it the hardest? Is there something more complicated about these grapes? Well, no, not really—it's just that there are a lot of them. Once we hit other regions and other red grape varieties, it's always simpler: there are usually only one or two major grapes involved in the blend, and more often than not, it's predominately a single grape that makes the wine. But Bordeaux has the complication of using two extremely well-known grape varieties (Cabernet Sauvignon and Merlot) as well as three lesser known varieties (Cabernet Franc, Petite Verdot, & Malbec) to make their wines. Since Bordeaux is a major wine region on the planet, and because Cabernet Sauvignon and Merlot are now widely produced across the planet, I felt it necessary to go for the gusto and tackle them all at once.

Cabernet Sauvignon is widely considered the king of red grapes, the master of the wine world. Tannic and bold and flavorful, it is usually thought to have great depth, complexity and age-ability. Much lesser known is that Cabernet Franc is the daddy of the king, genetically speaking, and is quite the court royalty on its own. And even though Merlot is widely mocked since the movie *Sideways* came out, it is virtually impossible to go to any wine store, wine restaurant, or wine-producing region on the planet that isn't offering up several versions of this grape. All of these guys are so widely produced, so widely recognized, and so widely consumed that I would like you to try an example of the Cabernet Sauvignon, the Merlot, and the Cabernet Franc as individual varietal wines, and then finish with a Bordeaux wine. Be forewarned: we are now treading on red wine ground; the wines will be heavier and more alcoholically potent, so watch yourself. For this exercise in particular, I'm asking you to try a lot of different heavy-hitting wines, so you may definitely want to do this with a group of friends. If you are determined to go solo, you should do these varietal wines one at a time, on different days, and then compare your notes later when you do the blended wine. Good luck, and may the Franc be with you…

What to grab:
- One bottle from each list of the following from your local wine shop or grocer:
  Cabernet Sauvignon
      Bottles $5-$15
        ➢ Bogle (California)
        ➢ McManis (California)

> Veramonte Reserva (Chile)
> Columbia Crest Grand Estate (Washington)
> Alamos (Argentina)
> Penfolds Rawson's Retreat (Australia)
> Geyser Peak (California)

Bottles$16-$30
> Robert Mondavi (California)
> Château Pipeau St.-Emilion (France)
> Bodega Catena Zapata Mendoza (Argentina)
> Toasted Head (California)

Bottles $30+
> L'Ecole Nº 41 Walla Walla Valley (Washington)
> Rubicon Estate CASK (California)
> Caymus Special Selection (California)

Merlot
Bottles $5-$15
> Concha Y Toro (Chile)
> Pine & Post (Washington)
> Blackstone (California)
> Columbia Crest Grand Estates (Washington)
> Finca Flichman (Argentina)
> Tommasi Le Prunee (Italy)

Bottles$16-$30
> Benziger (California)
> Ken Wright Cape Diamond (Oregon)
> Three Families (California)
> Sterling (California)

Bottles $30+
> L'Ecole Nº 41 Seven Hills Vineyard Walla Walla (Washington)
> Sterling Three Palms (California)
> Shafer Merlot (California)

Cabernet Franc
Bottles $5-$15
> Musaragno Organic (Italy)
> Ironside (California)
> Horton Cabernet Franc (Virginia)

Bottles$16-$30
> Mabilileau St. Nicolas de Bourgueil Loire Valley (France)
> Villa Appaliacia (Virginia)
> Domaine de la Noblaie Chinon les Chiens-Chiens (France)
> Conn Creek (California)

Bottles $30+
> Corte (California)
> Chappellet Vineyard Pritchard Hill Estate (California)
> Chateau Cheval Blanc 2000 (France)($1000) (bottle Miles drank in McDonalds at the end of *Sidways*)

A Bordeaux Blend
Bottles $5-$15
> Chateau La Moth de Barry (France)

> Chartron La Fleur Bordeaux (France)
> Abbaye De Saint Ferme Les Vignes Du Soir (France)
> Foris Fly Over Red (Oregon)

Bottles$16-$30
> Shotfire Ridge Barossa Cuvee (Australia)
> Chateau Cap de Faugeres Cotes de Castillon (France)
> Chateau Bujan 2004 Cotes de Bourg (France)

Bottles $30+
> Chateau Cantemerle (France)
> L'Ecole Nº 41 Seven Hills Perigee Walla Walla Valley (California)
> Chateau Cos d'Estournel (France)

Don't worry if you can't find these exact labels. You can always ask your local wine store salesperson to help you pick out a good representation of the varieties I have described above. Show them the list and tell them to get you something close to it.

What to do:

1. First open the bottle(s) and try immediately. What is the color, the aroma, the body, and the taste of each wine? Record all impressions. Write down some descriptions of your experience, both what you smell and how it tastes. You should be able to detect some subtle difference in flavors and aromas in these wines…and perhaps even different colors as well, but that is not of great significance for our exercise. Really work hard at picking out and describing differences between each of these varietal wines. As always, include as many descriptors as possible to define each wine, even if some may be a stretch. Check below for a bit of varietal vocabulary to choose from.

2. Next wait at least 2 hours for the wine to "open up," leaving the cork off the wine to allow oxygen to get in. Try the wine again. Has the smell changed? How so? What about the taste? Is it fruitier? Does it taste better to you? Really work hard at picking out and describing differences between each of these varietal wines. As always, include as many descriptors as possible to define each wine. As you go back and forth from one wine to the next a few times, be sure to cleanse your palate with a bit of bread or plain crackers. Sip some water as well, if you absolutely must. But don't eat any other food yet—it will easily alter the results of your descriptions. Don't worry: you can finish the bottles with dinner later.

3. Now let's revisit Lesson #1: try the wines with cheese or any appropriate food.   Record your reactions. Has anything changed in the taste? Does the wine seem smoother or lighter or fruitier? Repeat with a different type of food if available. Again, detect some differences and make yourself write down some different descriptors.

4. Finally (if possible), save a bit of wine in the bottle. Let it sit exposed (no cork) on the counter for a few more hours. Let it go overnight if you want. Then try it again on its own, without food. What's it like

now? Write it down. Compare this with your first impression of the
wine right after you opened it. What has happened?

5. If you purchased a Bordeaux blend wine, repeat steps 1-4 with it at
    another time, or another day depending on how much you have
    consumed already. Can you sense components of those grape
    varieties in this blended wine? And here is an extra assignment: go
    on-line or hit up the Vintage Cellar and find out what grapes and
    what % of each grape were used on the blend you tried. It's not that
    difficult to get this info, and each brand is slightly different.

6. Keep the rest of the bottles for dinner. Try with a variety of different
    foods. What works? What doesn't? What foods bring out more
    flavors in the wine? What foods overwhelm the wine? I want you
    to experiment as much as you want to see what works best for you,
    but listed on the next several pages are some variety/food pairings
    that do seem to bring out the best in these particular reds that you
    might want to try.

What to look for:

The whole point of this exercise, and others to follow, is to get you
fluid and fluent in recognizing and describing the major varietal wines
produced in the world today. Of course you know that the term *varietal
wine* means: a wine made predominately, if not exclusively, from one type
of grape. As you do the tasting, focus on comparing and contrasting these
different wines, with more of an emphasis on contrasting. However, for
this assignment we are also going to try a **blended wine** as well…and one

of the most famous blends in the world, the *Bordeaux blend*. Outside of Bordeaux you may see this blend referred to as Meritage.

Blended wines are those that are intentionally crafted from many different grapes to achieve a desired taste…they are not trying to emphasize the character of a single variety; it's all about the combination man! And I'll let you in on a little secret: blends were the standard of the past, and are the future! We will live to see blended wine expressions from all major production areas, as it used to be the way that everybody made wine. It is only in the current era that varietal wines have gained dominance—but I don't think they are going to hold the top slot for too much longer. Blending allows the winemaker to utilize all of his/her skills to craft a unique product, and also gives the versatility to make a good wine even in poor harvest years. Blends are the bomb!

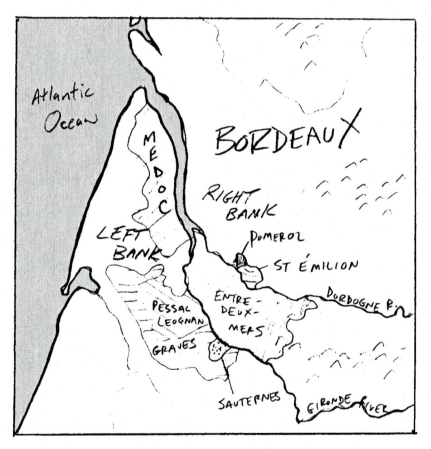

Well, now that you are done, what did you think? These reds may have been a bit trickier to tell apart than a lot of the whites were. Especially the Cabernet Sauvignon and the Cabernet Franc: they are closely related genetically, and you made not have been able to sense much difference between the two. Merlots are usually a bit more accessible and fruity, so that one may have been a bit easier to pick out of the crowd. And to be honest, I really don't expect you to be able to distinguish the different varietals in the blended wine: I'm more about you being familiar with what it represents as a blend than what is exactly in it.

Later exercises in this Boyer tasting round-up will compare some varietal wines from different regions, but you should do them on your own as often as possible too. When purchasing a bottle of one of these wines for dinner, go ahead and grab an inexpensive version of the same grape from a different part of the world. Your knowledge will grow exponentially, even if it is at the expense of your wallet decreasing mathematically. Old World Cabs and Merlots (if you can even find them as varietal wines) are radically different than their New World counterparts. Especially when comparing a classic old French wine to a big bomb California or Australia Cab. It can be a night and day experience. But I've become a bore; let's look at some specifics…

The Specifics:
Let's take a look at each of these major varieties one at a time to provide some tasting and smelling clues, some regional background, and also some food paring hints. Then we'll do the blend. Let's start with…

Cabernet Sauvignon
This grape is the king. The bomb. The bad-ass. The master of all lesser red grapes groveling before it. Okay, maybe it's not all that, but boy does it pack a punch! The extreme tannic component and pigment derived from the skins gives this wine lots of color, lots of mouth-pucker, and lots of durability. Cabernet's high tannic structure is a product of its thick-skinned, dark colored grapes which add tons of complexity to the already flavorful juice. This is a wine that can easily be crafted so intensely that it becomes a necessity to age it for a few years just to make it approachable. It is precisely because of this intensity that folks in Bordeaux have long understood the need to blend it with other grapes in order to smooth out the wine and make it more accessible.

Long considered the top of the heap of the classic wine grapes, Cabernet can be assertive and rough when young, and become intricately complex and velvety when softened with age. That's why some of the most expensive wines on the planet are typically aged Bordeaux's and high-end Cabs from other regions. Young or old, they are typically flavorful, dry, full-bodied, mouth-warming and intense. Herbaceous, cedar and cassis often make it into the list of descriptors of this wine, and below are a few more adjectives for you to consider…

Distinct Cabernet Sauvignon descriptors: Cedar, cassis, herbaceous, eucalyptus, bell pepper, sandalwood

General Cabernet Sauvignon descriptors: Berries, bell pepper, blackberry, blackcurrant, black olive, cedar, chocolate, cigar box, cinnamon, clove, coffee, cooked beetroot, dusty, earthy, eucalyptus, herbaceous, leafy, leather, licorice, menthol, mint, nutmeg, peppermint, perfumed, plum, soy, strewed rhubarb, tobacco leaf, tomato leaf, truffle, vegetative, violets, wild berries.

Where Cabernet Sauvignon is grown: Bordeaux region especially Medoc and Graves, northeastern Italy, California, Washington, Oregon, Texas, New York, all across Australia, Tuscany in central Italy, Penedes & Ribera del Duero in Spain, Chile, Argentina...oh hell, just about every place makes Cab anymore, but the above list has the longest track record.

What foods are often paired with Cabernet Sauvignon: Mild to strong to very strong cheeses, beef, grilled beef, roast beef, barbecue, game bird, wild game, wild boar, hearty pasta dishes, anything in a deep red sauce, any dark meat, spaghetti in a serious red sauce, steaks...oh yes! Steaks! A rare steak on the grill with blue cheese crumbled on top and a deep Cab....mmmmmm. And chocolate...yes! The darker the better. Fudge, chocolate cake...etc... You know what I'm saying!

And let's do some specific cheeses too
Definitely do: 2 Year Cheddar, Dry Jack, Manchego
Will work: Lake's Edge, Marrisa, Pont-L'Eveque, Romano Pradera
Possibly try: Camembert, Sharp Cheddar, Danish Blue, Grafton Village Cheddar, Cambozola

Merlot
The most overblown grape on the planet. Overproduced, over-consumed, and the wine grape that I'm most over. And it ain't over yet. So much planting of this grape has occurred to meet market demand that it will be years before the tidal wave of Merlot starts to subside on the market. But the flood waters have just started to recede. For those of you not in the know, its why that dude in *Sideways* was ridiculing Merlot so much... remember that classic line when he says "Okay I'll do it, but no Merlot! I won't drink Merlot!" This is one of the immortal blending grapes of Bordeaux, and just that: a blending grape. The idea that this grape would ever become one of the most produced varietal wines would have been laughable 50 years ago. It's just not up for the challenge. (With some exceptions: see Pomerol, Chateau Petrus, & St Emilion in Bordeaux which are overwhelmingly Merlot and more overwhelmingly priced.)

Having just said that, there are some really awesome ones out there! And it's precisely because so many folks are producing so much of the stuff, that some winemaker's have taken the ultra-high road and decided to craft some exceptional wines out of Merlot. There are also some excellent mid-range expressions of this varietal wine too...with particular emphasis on *some*. But make no bones about it: an ocean of mediocre Merlot awaits you at every restaurant, wine shop and grocery store. How did this happen?

I just described Cabernet Sauvignon to you as bold, assertive, and typically complex. All awesome characteristics for a wine to have when being consumed by experienced wine drinkers. However, our country as a whole is a wine-consuming virgin, and as such needs something much easier on the palate to start off with. Consider Merlot like a set of training wheels on your red-wine drinking bike: it makes things easier and props you up as you start down the path of serious red wines. Merlot is often described as fruity, soft, subtle, or supple because its tannic component is nowhere near as aggressive as Cab. As such, its easy-drinking and approachable and rounded even at a young age. It also is an easy pairing for most foods as well as being easy-sipping on its own, which is why it has been so embraced by the American public.

And our country has embraced it big time: it has become the easy-drinking red wine rage for the last couple of decades. And as America goes, the world goes. Every single country is producing Merlots, and labeling them as Merlots, mostly for the American market. I've seen bottles labeled Merlot from every single wine-producing country. Sheesh! Even the French are exporting low-end bottles labeled Merlot! That would have been considered blasphemy five years ago! I think it still is a blasphemy…

But here is the good news: fads come and go, and Merlot is on its way out. Not the really nice, well-crafted versions mind you, but the ocean-tanker mediocre ones. Now that America is finding its 'wine legs', and there are so many other wine options available, Merlot consumption will probably start to tail off. It will resume its role as an excellent blending grape, particularly for the Bordeaux blend.

Distinct Merlot descriptors: Blackberry, plum, cassis, fruity

General Merlot descriptors: Aromatic, bacon, beet, blackberry, blackcurrant, black olive, caramel, cherries, chocolate, cinnamon, coffee, earthy, fruitcake, herbaceous, leafy, leather, meaty, mulberry, musk, perfumed, plum, raspberry, redcurrant, rhubarb, roses, spicy, stewed plums, tobacco leaf, truffle, underbrush, violets

Where Merlot is grown: Everywhere…i.e. see Planet Earth. But the classic growing areas have been Bordeaux, especially Pomerol and St Emilion, and Northeastern Italy. California, Australia, Argentina, and Washington do some serious work with Merlots now too, but I was being serious about the whole world: you can find them from Hungary to New Zealand to Chile to China.

What foods are often paired with Merlot: Merlot is really a catch-all for food. Its hard to go wrong. But I know these things work from experience: Mild to strong cheeses, pork, grilled pork, pork loin, pork chops, beef brisket, beef stew, roast beef, beef pot-roast, steak, pepper steak, anything in a light red sauce (tomato or fruit based), pretty much any desert that is chocolate based. Those deserts really bring out the fruit in the wine and make for a pleasant experience…as if eating desert wasn't pleasant enough.

And let's do some specific cheeses too
Definitely do: Aged Provolone, Aged Gouda, Azeitao
Will work: Garrotxa, English Cheddar, Camembert, Taleggio
Possibly try: Brie, Pont L'Eveque, Roaring 40's Blue, Buffalo Mozzarella

Cabernet Franc

Cabernet Franc is as under-rated as Merlot is over-rated. Related to Cabernet Sauvignon, Franc has a less intense flavor profile and is a bit thinner-skinned than its kin, and that translates to less tannins, slightly less color, and slightly less body. But what it does have is good acidity and some extremely distinct aromas and flavors that are sometimes described as minerally, ashy, smoky, or briar-like. These characteristics make it prized as a blending grape in Bordeaux and increasingly abroad. However, I believe that my good friend Franc is likely to become a popular varietal wine as well. You can find some very interesting varietal Cabernet Francs out of the Loire Valley, New Zealand, California, and actually right here in Virginia too! It is that mysterious, funky smoky flavor that makes it so intriguing, and ultimately may serve to increase its visibility as a stand-alone wine character.

Distinct Cabernet Franc descriptors: Cedar, smoke, briar (a kind of woody/herbaceous sensation that is distinctly not oak), pencil shavings, black olive, raspberry.

General Cabernet Franc descriptors: Blackcurrant, black olive, dusty, earthy, gooseberry, herbs, leafy, licorice, mint, musk, nutmeg, pepper, plum, raspberry, sandalwood, savory, spicy, violets

Where Cabernet Franc is grown: Bordeaux and Loire Valley in France, northeastern Italy, California, and Washington and lo and behold!: the great state of Virginia is turning up as a major producer of this varietal wine. (By the way, this is not self-promotion: I am not a spokesperson for Virginia wine. Just try the Franc from Horton Vineyards and tell me that its not good—I dare you!)

What foods are often paired with Cabernet Franc: mild to strong cheeses, lamb (*very popular pairing), grilled pork, game birds, wild game, anything gamey; game on, venison, duck, goose, heartier red sauce pasta dishes, a solid good pizza. Especially a veggie pizza.

And let's do some specific cheeses too
Definitely do: Emmenthaler, Fontina, Havarti
Will work: Port Salut, Raclette, Feta
Possibly try: Brie (mild), Camembert, Cheddar (sharp)

Additional barrel flavors you may detect in any of the wines depending upon how much time it spent in oak: (From Lesson 4: Got Wood?)

➤ *Specific to these red wines*: Almond, bacon, burnt, cashew, cedar, coconut, nutty, pencil shavings, sawdust, smoky, spicy, toast, toffee,

vanilla
- ➤ Woody descriptors: Cedar, charred wood, cigar box, coffee, green wood, oaky, pencil shavings, resinous, sandalwood, sawdust, smoky, toasty, tree bark, vanilla.
- ➤ Barrel flavor (malolactic fermentation) descriptors: Almond, banana, burnt caramel, butter, buttered popcorn, butterscotch, cashew, coconut, lanolin, vegemite
- ➤ Lees flavor descriptors: Baked bread, creamy, cheesy, leesy, nutty, yeasty, yogurt

One final note on these wines:
Try to focus on the fruit components as you do this exercise, knowing full well that everything else changes according to region, climate, and winemaker influence. What have you learned? Hopefully you picked up some differences in these three varieties, but more importantly that you now know what Bordeaux wine consists of, as well as all those other wines around the world which mimic the Bordeaux recipe. And the individual varietal wines can be radically different depending upon region of production: some are soft and supple, others are powerhouses.

At the beginning of the lesson I told you that Bordeaux uses five grape varieties (Cabernet Sauvignon, Merlot, Cabernet Franc, Petite Verdot, & Malbec) but we only covered three of them. That is because they are the three big boys of the blend in terms of how much of each grape is in there. Cabernet Sauvignon and Merlot represent the lion's share of acreage and of the grape juice used in Bordeaux, while Franc usually constitutes a very distant third. Petite Verdot and Malbec make it into some Bordeaux wines in trace amounts, so I didn't want to deal with them here (we will pick up Malbec's story in a latter lesson). Merlot actually is the leader in terms of acreage and production. It may best be considered the base for the wine to be blended on. Sauvignon is such a powerhouse that Merlot is added to it (in greater or lesser amounts) in order to soften it up and round it out. Franc and others are added for specific flavors and acids.
The end product is something distinctively Bordeaux: Sauv has the power, Merlot has the suppleness, Franc adds some flavor. How delightful!

But lastly know this: there is no single standard recipe for red wines of Bordeaux. Some wines have more Cabernet Sauvignon (Medoc & Graves, referred to as the *Left Bank* of Bordeaux); some wines have more

Merlot (St. Emilion, Pomerol on the *Right Bank* of Bordeaux); some wines are 100 % Merlot (Pomerol); a lot of wines have percentages of all five grapes; some only have three grapes. That's the beauty of blended wines: they are all about crafting the best product that the terroir, the harvest, and the winemaker have to offer instead of focusing on just bringing out the characteristics of a single grape. Think about it. That kind of makes sense, doesn't it?

	Comments
Cabernet Sauvignon	
Color:	
Bouquet (smell):	
Taste:	
Body:	
Compare/Contrast to other wines:	
Merlot	
Color:	
Bouquet (smell):	
Taste:	
Body:	
Compare/Contast to other wines:	
Cabernet Franc	
Color	
Bouquet (smell):	
Taste:	
Body:	
Compare/Contrast to other wines:	

Lesson 12: Burgundy's Big Hitter: Pinot Noir

What it smells like. What it tastes like. Where you get them from. What foods bring out the best in them. And what the hell is all the fuss about when it comes to this grape and its wine?

To start, Burgundy = Pinot Noir. If it's red, and the label says Burgundy, or Bourgogne in French, then its 100% Pinot Noir. Learn it. Live it. Love it.

If you have been doing these lessons in order, then you just finished tasting the Bordeaux trio which contained some powerhouse red varieties. We are definitely switching gears with this tasting in a couple of different ways: not only will we be doing just a single wine, but this varietal wine is known more for its delicateness and finesse as opposed to any powerhouse flavors or tannic structure. But don't be fooled: one of the reasons that Pinot Noir gets so much attention is that it can be crafted into something so complex and so exceptional that it has become something on the order of a 'Holy Grail' that wine imbibers search for like it's a deity-inspired quest.

As such, it has created a cult of personality around itself that is second to none. And the impetus for winemakers to tap into this cult has resulted in an extremely bipolar dichotomy when it comes to Pinot Noir wines: the exceptional top tier of fantastic wines, and a huge number of bottom feeders that are flabby and unexciting. If I had to pick a single wine which really has absolutely no middle ground, this would be it. Many shoppers ask me all the time, "Hey, I'm looking for a really, really good Pinot Noir for under $20,'" to which I respond: "Hell yeah, so am I!" Because they just don't exist. Or if they do, they are well hidden. If you want a true representation of the best this grape has to offer, you are going to have to pony up the bucks. Why is this?

Go ahead and grab a bottle or two, pull the cork, pour a glass, and I'll explain it to you below...

What to grab:
- One bottle (or 2 or 3 if you are going to compare across price spectrum) of the following from your local wine shop or grocer:
  Pinot Noir
    Bottles $5-$15
    - ➢ Mirassou (California)
    - ➢ Silverwing (Australia)
    - ➢ Acacia (California)
    - ➢ Castle Rock (Oregon)
    - ➢ Domaine Chevillon-Chezeaux (France)
    - ➢ Babich Pinot Noir 2005 (New Zealand)

Bottles $16-$30
> ➢ Stephen Vincent (California)
> ➢ La Crema (California)
> ➢ Les Penitents Alphonse Mellot (France)

Bottles $30+
> ➢ **Any bottle from Patricia Green (Oregon) $35 - $55
> ➢ **Any bottle from Ken Wright (Oregon) $35- $80
> ➢ **Any bottle from St. Innocent Vineyards (Oregon) $45 - $85
> ➢ Louis Jadot Les Beaux Monts Vosne Romanee (France)

Don't worry if you can't find these exact labels. You can always ask your local wine store salesperson to help you pick out a good representation of the varieties I have described above. Show them the list and tell them to get you something close to it.

What to do:

1. First, open the bottle(s) and try immediately. What is the color, the aroma, the body, and the taste of each wine? Record all impressions. Write down some descriptions of your experience, both what you smell and how it tastes. As always, include as many descriptors as possible to define the wine, even if some may be a stretch. Check below for a bit of varietal vocabulary to choose from.

2. If you choose to buy more than one bottle of Pinot Noir to compare the quality/price scenario, you should be able to detect some subtle difference in flavors and aromas and smoothness between your wines…and perhaps even different colors as well, but that is not of great significance for our exercise. Really work hard at picking out and describing both the similarities and differences between the two wines. As always, include as many descriptors as possible to define each wine, even if some may be a stretch. Check below for a bit of varietal vocabulary to choose from.

3. Next wait at least 2 hours for the wine to "open up," leaving the cork off the wine to allow oxygen to get in. Try the wine again. Has the smell changed? How so? What about the taste? Is it fruitier? Does it taste better to you? As you go back and forth from one wine to the next a few times, be sure to cleanse your palate with a bit of bread or plain crackers. Sip some water as well, if you absolutely must. But don't eat any other food yet—it will easily alter the results of your descriptions. Don't worry: you can finish the bottles with dinner later.

4. Now let's revisit Lesson #1: try the wines with cheese or any appropriate food.   Record your reactions. Has anything changed in the taste? Does the wine seem smoother or lighter or fruitier? Repeat with a different type of food if available. Again, detect some differences and make yourself write down some different descriptors.

5. Finally (if possible), save a bit of wine in the bottle. Let it sit exposed (no cork) on the counter for a few more hours. Let it go overnight if you want. Then try it again on its own, without food. What's it like

now? Write it down. Compare this with your first impression of the wine right after you opened it. What has happened?

6. Keep the rest of the bottles for dinner. Try with a variety of different foods. What works? What doesn't? What foods bring out more flavors in the wine? What foods overwhelm the wine? I want you to experiment as much as you want to see what works best for you, but listed on the next several pages are some variety/food pairings that do seem to bring out the best in these particular reds that you might want to try.

What to look for:

The whole point of this exercise, and others to follow, is to get you fluid and fluent in recognizing and describing the major varietal wines produced in the world today. Of course you know that the term *varietal wine* means: a wine made predominately, if not exclusively, from one type of grape. As you do the tasting, focus on comparing and contrasting these different wines, with more of an emphasis on contrasting. However, for this assignment we are only doing a single variety, but if you have the money and interest, try a low $ Pinot, a mid-range $$$ Pinot, and a high-end $$$$$ Pinot, and compare/contrast what happens with increase in price.

No matter what region you get the wine from, and no matter how much money you spend, you should be able to pick out the unique fruit and earthiness characteristics of this grape by the end of this exercise. Pinot Noir is almost always found as a varietal wine, no matter where it's from— even in Europe it is rarely used as a blending grape. I can think of one big exception to that rule, which is Champagne: Pinot Noir is one-third of the trifecta of grapes used in the blend for the famous fizzy stuff. But because Pinot Noir is a fussy grape to grow, and a fussy wine to craft, most winemakers either go for the gusto and make the varietal wine, or just stay away from it altogether.

Later exercises in this Boyer tasting round-up will compare some varietal wines from different regions, but you should do them on your own as often as possible too. When purchasing a bottle of one of these wines for dinner, go ahead and grab an inexpensive version of the same grape from a different part of the world. Your knowledge will grow exponentially, even if it is at the expense of your wallet decreasing mathematically. Old World Pinot Noirs are radically different than their New World counterparts. As with many other varietals, the Old World stresses a more acidic backbone which devolves in time to create an exceptionally smooth and intricate wine while the New World Pinots are most often focused on being fruit-forward and quaffable. But even those New World Pinots can be exceptionally complex when done right. Its just something about the grape… What something? Well, let's take a look….

The Specifics:

Let's take a closer look at this elusive grape, made into a much-hyped wine, which is sometimes worth all the hype. Here are some tasting and smelling clues, some regional background, and also some food paring hints.

Pinot Noir

As I've already suggested, with Pinot Noir when it's good, it's good. And when it's great, it's to die for. Bordeaux varieties may be the world workhorses, and the New World has made some titans of other varieties, but Pinot Noir stands alone as a wine whose definitive characteristic is that of complexity, finesse, and silkiness…all simultaneously. Of course, that is when it is at its best…which doesn't necessarily happen that often. Pinot Noir's complexity of body and sensations is the reason that the descriptor list for its aromas and flavors and textures can go on forever.

Very often it asserts its fruitiness up front, has some vegetal/spice in the

middle, and its earthy character on the finish. Just look at the massive descriptor list I've given you to work with; it says it all. There is simply a lot going on in these wines. Somehow the wines are light, yet full-bodied and rich. Full-flavored, but not heavily alcoholic. Not necessarily tannic or acidic or fruity or vegetal or spicy…but somehow a little of it all. Delicate is the operative word with these wines. Delicate, soft, silky, velvety textures and tastes. Sounds just too titillating, doesn't it? But again to re-stress: that is when it's crafted very, very well. Which isn't often. And why is that?

Well, to put it bluntly, this grape is a bitch. It is temperamental, which means it's a whiner. It is prone to rapid genetic variation, which means it won't stay put. It's a thin-skinned grape, which means it's a wuss. It ferments erratically, which means it can't easily be controlled. And even after all is said and done, it's not the longest lasting thing in the cellar, which means it's a weakling.

Whiner, wuss, weakling….what? Yep. Its Pinot.

A bit more elaboration: Pinot Noir is extremely prone to destruction via the elements. Too hot, too dry, too cold, too wet, and it just doesn't produce great fruit. It even buds very early which means it's prone to spring frost. The grape is just a whiner about everything! Everything has to be just right! It also mutates rapidly, which means the offspring can be radically different than the parent plant. It also means that when planters take 20 cuttings of the exact same plant to 20 different places in the world, the fruit can become radically different as the plant rapidly adapts to unique conditions. Now, that can be a good thing if it does well, but I think you are starting to get the picture that this scenario of 'doing well' doesn't happen that often.

Pinot is also a physically thin-skinned grape, which makes it more vulnerable to just about every single grape and grapevine disease on the planet. (It's also why Pinot wines are typically lighter in color.) Bacteria, viruses, rot, and even insects seem to have a natural knack in destroying Pinot Noirs. And it's easy too! To add insult to injury, even when a grower successfully pulls off a Pinot Noir crop, the juice sometimes ferments so erratically that it appears to be on the brink of explosion, and that is not something which makes for a desirable situation to extract the proper delicate flavors. And after bottling, don't expect them to age indefinitely: Pinot Noirs typically have half the cellar life of a big beefy Bordeaux or California red.

So now you know what the fuss is about. It's a total bitch grape that in some special circumstances can become Cinderella. There are those that believe that all the time and pain and suffering are worthwhile once you experience the real manifestation of greatness. Well, I don't know about all that, but I can teach you this: any commodity that is scarce commands a higher price. And great Pinots will have to be paid for....

But get to it, and do it! One last hint: Old World vs. New World Pinot Noirs are not only different, but that difference can be significant. So significant that we will do a comparative tasting between the two in a later lesson. Just know this for now while you look over the descriptor list:
- Old World Pinots stress the earthiness more: earthy, leather, musty, barnyard, mushroom
- New World Pinots stress the fruit more: cherry, raspberry, plum, violets

Consider where your Pinot is from as you search for descriptions during your tasting. Good luck!

Distinct Pinot Noir descriptors: Cherry, raspberry, tomato, cedar, mushroom, earth, earthiness, barnyard, truffle, leather, saddle leather

General Pinot Noir descriptors: Cherry, strawberry, raspberry, blackberry, black cherry, red cherry, plum, stewed plums, strawberry, black currant, black pepper, black tea, cranberry, dried flowers, fruity, lavender, licorice, violet, very subtle rose petal, sassafras, rosemary, cinnamon, rhubarb, beet, oregano, ripe tomato, green tomato, tomato leaf, tobacco leaf, mint, musk, perfumed, green tea, black olive, mushroom, earth, barnyard, truffle, forest floor, bacon fat, dark chocolate, gamey, mocha, humus, prune, tree moss, leather, meaty, cedar, cigar box, hay, vanilla, oak, smoke, toast, tar, coconut, wild berries.

Where Pinot Noir is grown: The original and still king of Pinot Noir is Burgundy, specifically a region called the Côte d'Or ("Slope of Gold"). Let me re-state that which I have said before: red Burgundy = Pinot Noir. But you can also find growers from across the planet experimenting with this variety, and don't be surprised to see it represented in virtually all major wine regions of the planet. It does seem to have an affinity for a slightly cooler climate, so some of the most promising Pinot places in the New World include the Willamette Valley in Oregon, and New Zealand. Look for these specific appellations in California too for your best bets for this grape: Santa Maria Valley, Russian River Valley, *Carneros, Anderson Valley, the Pinnacles, and the Santa Lucia Highlands.

What foods are often paired with Pinot Noir:
Salmon, smoked salmon, grilled salmon, and salmon any damn way you want it! Tuna steaks! Yes! Also, mild cheeses, beef, grilled beef, roast beef, barbecue, game bird, wild game, quail rocks out to Pinot, chicken, chicken cooked in red sauce especially, roasted lamb, braised lamb, pheasant, duck, hearty pasta dishes, anything in a light red sauce, mushrooms, and actually any dish that has mushroom as a flavor or as a main element is great, pork, veal, ham, and a perennial American pairing is Pinot and turkey. And beef stew or any meat stew…particularly one called Cassoulet. And don't forget the wine's namesake dish: Beef Bourgogne

- ✓ **Beef Bourguignon** (French: Bœuf bourguignon) is a well-known, traditional French recipe. It is essentially a type of beef stew prepared with cubed beef stewed in red wine (preferably an assertive, full-bodied wine such as Burgundy), generally flavored with garlic, onions, carrots, lardons, and a bouquet garni, and garnished with pearl onions and mushrooms.
- ✓ **Cassoulet** is a rich, slow-cooked bean stew or casserole originating in the southwest of France, containing meat (typically pork sausages, pork, goose, duck leg and sometimes mutton), pork skin (couennes) and white haricot beans

And let's do some specific cheeses too
Definitely do: Comte, Eppoisses, Caerphilly
Will work: Clothbound Cheddar, Farmhouse Cheddar, Vacherin Fribourgeois, Adrahan
Possibly try: Zamorano, Swiss, Gouda, Cheddar (mild), Feta, Raclette

<u>**Additional barrel flavors**</u> you may detect in any of the wines depending upon how much time it spent in oak: (From Lesson 4: Got Wood?)
- ➢ Woody descriptors: Cedar, charred wood, cigar box, oaky, sandalwood, sawdust, smoky, toasty, vanilla, coconut, sweet wood, tar.
- ➢ Barrel flavor (malolactic fermentation) descriptors: Almond, banana, burnt caramel, butter, buttered popcorn, butterscotch, cashew, coconut, lanolin, vegemite (these would be very light in Pinot Noirs, if detectable at all)
- ➢ Lees flavor descriptors: Baked bread, creamy, cheesy, leesy, nutty, yeasty, yogurt (these would be rare in Pinot Noirs)

One final note on Pinot Noir….

Try to focus on the fruit components as you do this exercise, knowing full well that everything else changes according to region, climate, and winemaker influence. And change they will: Pinot Noirs in particular can

be radically different depending mostly upon the climate and/or particular harvest season of each region. Unlike some other varieties, a bad year for the grapes almost always makes for a bland or outright un-fun wine experience. Why? It's a temperamental grape; which makes a very delicate wine; a wine which is unblended, making it a sole expression of the vintage—all these factors together make it very difficult to pull off.

But when someone does pull it off…wow. It can be a spectacular experience for the palate. Pinot Noir has this unique capacity to be delicate, light, fruity, earthy, sensuous, velvety, and marvelously complex all at once. No other wine seems to simultaneously provide these extremes of body and flavor, and perhaps that is what is so intriguing about them. Maybe that's why wine connoisseurs seek them out; perhaps why some are willing to spend tremendous amounts of money on them to experience this uniqueness. Go rent _Sideways_ again and listen to the great conversation between Miles and Maya about the complexities of this wine, and you will begin to see why some are so passionate about them.

And I'm all for passion my friends…but make no bones about it: passion always comes at a price. If you want to experience the best Pinot Noir has to offer, the price will be high. This is a personal call on my part, but Pinot Noir is definitely one of those grapes/wines that the more you pay, the more you can expect. The inverse holds true as well: don't expect to get your socks knocked off with a $10 Pinot, 'cause its just not going to happen. Maybe not even for $50…maybe not even for a $100. But someday you will cross one of these truly magical wines and then, and only then, will you see what all the fuss is about.

But don't hold your breathe waiting for that day to dawn. Get out there and drink more wines. Go forward to the next lesson, as we start to talk about some Italian wine magic…

	Comments
Pinot Noir $	
Color:	
Bouquet (smell):	
Taste:	
Body:	
Compare/Contrast to other wines:	
Pinot Noir $$$	
Color:	
Bouquet (smell):	
Taste:	
Body:	
Compare/Contrast to other wines:	
Pinot Noir $$$$$	
Color	
Bouquet (smell):	
Taste:	
Body:	
Compare/Contrast to other wines:	

Lesson 13: The Italian Stallions: Sangiovese, Nebbiolo, & Barbera

What they smell like. What they taste like. Where you get them from. What foods bring out the best in them. And many of the aliases that they go by. And why the hell do they go by so many names? Barolo, Barbaresco, Morellino di Scansano, Chianti Classico, Vino Nobile de Montepulciano, Brunello di Montalcino...so many names, for just a couple of grapes! What gives?

We have talked about a lot of grape varieties from a lot of different places, and those same varieties are being planted and experimented with all over the world. As I have suggested in past lessons, you can get a Merlot or a Cab or Chardonnay from any corner of the globe. But now we get to some grapes that are distinctly Italian. Some would say imminently Italian. Some would say intimately Italian. And man oh man oh man, do I dig these Italian grapes and their wines, as do so many across the planet. Indeed, in some of their incarnations Nebbiolo and Sangiovese are considered the 'Kings of the Wine World'!

But don't get too excited yet: those referred to as 'royalty' are not ones that you will be bumping into for this introductory lesson on these varietals—unless you have a whole lot of expendable capital laying around that you want to drink up! These stallions are not just the workhorses of the most important Italian red wine producing regions, but they can also be crafted into complex thoroughbreds which qualify them for their regal reputation. And even though it's not as noble, Barbera is a grape that I want you to be familiar with as well since it seems to be making headway as an increasingly produced, and increasingly recognized, varietal wine in its own right.

Mama Mia! Vino Bellissimo! Vino Bellissimo! Let's taste these Italian Stallions so you can see for yourself!

What to grab:
- One bottle from each list of the following from your local wine shop or grocer (or just do one at a time at your own pace...because these guys can get expensive, fast):

Sangiovese

Bottles $5-$15
  - ➤ Monte Degli Angeli Sangiovese Puglia (Italy)
  - ➤ Fossi Chiante Classico (Italy)
  - ➤ Dievole Dievolino (Italy)
  - ➤ Bolla Sangiovese Di Romagna (Italy)

Bottles $16-$30
  - ➤ Corte Alla Flora Rosso Montepulciano (Italy)
  - ➤ Benegas (Argentina)
  - ➤ Corte alla Flora Vino Nobile di Montepulciano (Italy)

Bottles $30+
- ➢ Fontodi Chiante Classico (Italy)
- ➢ Verbena Brunello (Italy)
- ➢ Lisini Brunello di Montalcino (Italy)
- ➢ Casanova Di Neri Tenuta Nuova Brunello Di Montalcino (Italy)

Nebbiolo
Bottles $5-$15
- ➢ Cantalupo Il Mimo Rosato Di Nebbiolo (Italy)
- ➢ Paolo Scavino Rosso Vino Da Tavola (Italy)

Bottles$16-$30
- ➢ Vdt Antonio Vallana 'Campi Raudii' NV (Italy)
- ➢ Lodali Barbaresco (Italy)
- ➢ Lodali Barolo (Italy)

Bottles $30+
- ➢ Barboursville (Virginia)
- ➢ Grimaldi Barolo (Italy)
- ➢ Ciabot Mentin Ginestra Clerico (Italy)

Barbera
Bottles $5-$15
- ➢ Oddero Barbera d'Alba (Italy)
- ➢ Sant Agata Alta Barbera d'Asti (Italy)
- ➢ Renwood Barbera 2005 (California)

Bottles$16-$30
- ➢ Martinetti Bric Dei Banditi Barbera d'Asti (Italy)
- ➢ Camerano Barbera d'alba (Italy)

Bottles $30+
- ➢ La Spinetta Barbera d'Alba (Italy)

Don't worry if you can't find these exact labels. You can always ask your local wine store salesperson to help you pick out a good representation of the varieties I have described above. Show them the list and tell them to get you something close to it.

What to do:
1. First open the bottle(s) and try immediately. What is the color, the aroma, the body, and the taste of each wine? Record all impressions. Write down some descriptions of your experience, both what you smell and how it tastes. You should be able to detect some subtle difference in flavors and aromas in these wines…and perhaps even different colors as well, and this IS of some significance for this exercise (the first time I've said that). Really work hard at picking out and describing differences between each of these varietal wines. As always, include as many descriptors as possible to define each wine, even if some may be a stretch. Check below for a bit of varietal vocabulary to choose from.
2. Next wait at least 2 hours for the wine to "open up," leaving the cork off the wine to allow oxygen to get in. Try the wine again. Has the smell changed? How so? What about the taste? Is it fruitier? Does it taste better to you? As you go back and forth from one wine to

the next a few times, be sure to cleanse your palate with a bit of bread or plain crackers. Sip some water as well, if you absolutely must. But don't eat any other food yet—it will easily alter the results of your descriptions. Don't worry: you can finish the bottles with dinner later.

3. Now let's revisit Lesson #1: try the wines with cheese or any appropriate food. Record your reactions. Has anything changed in the taste? Does the wine seem smoother or lighter or fruitier? Repeat with a different type of food if available. Again, detect some differences and make yourself write down some different descriptors.

4. Finally (if possible), save a bit of wine in the bottle. Let it sit exposed (no cork) on the counter for a few more hours. Let it go overnight if you want. Then try it again on its own, without food. What's it like now? Write it down. Compare this with your first impression of the wine right after you opened it. What has happened?

5. Keep the rest of the bottles for dinner. Try with a variety of different foods. What works? What doesn't? What foods bring out more flavors in the wine? What foods overwhelm the wine? I want you to experiment as much as you want to see what works best for you, but listed on the next several pages are some variety/food pairings that do seem to bring out the best in these particular reds that you might want to try.

What to look for:

The whole point of this exercise, and others to follow, is to get you fluid and fluent in recognizing and describing the major varietal wines produced in the world today. Of course you know that the term *varietal wine* means: a wine made predominately, if not exclusively, from one type of grape. As you do the tasting, focus on comparing and contrasting these different wines, with more of an emphasis on contrasting.

Later exercises in this Boyer tasting round-up will compare some varietal wines from different regions, but you should do them on your own as often as possible too. When purchasing a bottle of one of these wines for dinner, go ahead and grab an inexpensive version of the same grape from a different part of the world. Your knowledge will grow exponentially, even if it is at the expense of your wallet decreasing mathematically. This can be very challenging with the Italian Stallions, because these grapes are not widely grown outside of Italy. You are starting to see some versions of Sangiovese and Barbera pop up in California and elsewhere, but Nebbiolo is almost entirely still an Italian affair. Que bella! Perhaps as it should be: geography still matters—at least for that grape.

The Specifics:

Let's take a look at each of these major varieties one at a time to provide some tasting and smelling clues, some regional background, and also some food paring hints. Let's start with…

Sangiovese

Maybe you've never even heard of this grape, but I'll bet you know it by another name: Chianti. Once again: Momma Mia! That's Italian! You mean Chianti, that cheap-o wine in the straw-covered bottle that you get at the Italian restaurant, and then you take the bottle home and put a candle in it—how romantic! Yep: that's the lowest, lowest, lowest-end version of a Sangiovese that you may have already tried. Fine Italian wine has had a lot of trouble establishing itself in the American market precisely because of the bad reputation that was heaped on it due to those mass-marketed Chiantis of the 1960s-1970s.

But oh, it's so much more! Sangiovese is the foundation block of Tuscany, and Tuscany itself is as close to a little piece of heaven that you will find on this earth! Sangiovese makes up anywhere from 80 to 90 to close to 100% of the blend in the entire Chianti region—the biggest and most recognized subset of Tuscany proper. Sangiovese accounts for 10% of all the grapes planted in Italy. It produces medium-bodied wines with classic prune, cherry character aromatics and significant acidity.

And a lot hinges on that acidity: Sangiovese's performance can be extremely diverse—from the insipidly thin, ascetic table wines of the most basic of Chiantis to the awesome heights of rich, concentrated, and complex reds like Brunello di Montalcino, Vino Nobile di Montepulciano or the rustic old fashioned charm of Morellino di Scansano. Why so much diversity?

Well, there is always the role, and goal, of the winemaker and what they are trying to craft. That accounts for some of the variability. But with this grape, it's more about the damn clones. Cloning? What? Why hasn't President Bush stopped this madness yet? Because this type of cloning happens quite naturally, although humans can manipulate the process to their advantage as well. As with most grapes, the plant material morphs slightly when taken to different locations, producing slight variations in the fruit. And no grape seed is 100% genetically true to its parent. So over the years, many variations on the plant develop. Sangiovese is perhaps the best example of how, within a single variety, many clones of varying capabilities can exist. Now you know why that mega-tons of Sangiovese go into making cheap Chiantis, while you have to pay mega-bucks to get a Sangiovese from the town of Montalcino—they are using two different clones of the same grape that have two distinct variations on the main theme. Montalcino thinks their clone is so special that they give it a unique name: *Brunello*. Once you try it, you will find out why it's so special….and friends, it is special….mmmmmmm…wow. But I digress…

However, there are some commonalities of all Sangioveses: because the grape is thin-skinned and is slow to ripen, they are almost always lightly-pigmented (with the *Brunello* clone being a slight exception). The wines, whether young or old, often cast a slight orange tinge at the edges that you can see as you tilt the glass. That stiff-backboned acidity that I have alluded to many times is also another key component of these wines. The nose on Sangiovese can also be tricky to identify distinctly, but often includes prunes, strawberry, violet, plum, dark cherry and earthy barnyard.

Distinct Sangiovese descriptors: strawberry, blueberry, orange peel, plum, violet, cinnamon, clove, thyme, vanilla, sweet wood, oak, smoke, toast, tar

General Sangiovese descriptors:
Blackberry, blueberry, black cherries, strawberry, capers, chocolate, dusty, earthy, herbal, leather, licorice, meaty, pepper, plum, prune, raspberry, sour cherry, spicy, tar, tobacco, and woody descriptors include almost, cedar, nutty, savory

Where Sangiovese is grown: The Tuscan specialty. Tuscany is in Central Italy north of Rome. They've been growing this stuff forever, and they grow the most. California is the only other place that is growing enough acreage to constitute a serious attempt at making these wines, and they are getting better at it as of late…

What foods are often paired with Sangiovese: Mild to strong to very strong cheeses, seafood in heavier cream sauces, particularly if they are red or brown. Salmon, poultry, game bird, heartier pastas, beef dishes, barbecues, duck, goose, grilled pork, Italian sausage, pork roast, lamb, cold cuts, a solid good pizza. Oh hell, just about everything listed on the menu at an Italian restaurant…if its got red sauce in it, on it, or beside it, have a damn Chianti with it. You can't go wrong. That's why Chianti is so definitively Italian!!!! And if you ever visit Tuscany, go with a local wine and wild boar—it's to die for…luckily, the boar will do that for you.

And let's do some specific cheeses too
Definitely do: Boschetto al Tartufo Bianchetto, Mozzarella
Will work: Taleggio, Garrotxa, Parmesan Reggiano, Pecorino, Provolone
Possibly try: Aged Gouda, Idiazabal, Serra, Gorgonzola (Dolce), Ardrahan

Nebbiolo

The King is here! And I'm not talking about Elvis! If you want to pay the most money for Italian wine, head to the Piedmont region of the northwest and look no further than Barolo and Barbaresco—made from 100% Nebbiolo. This inky, black-as-night-skinned variety is responsible for some of the biggest, best-est, and most age-able red wines of Italy...and the world. When young, these wines can be huge, full-bodied, massively tannic, acidic, and bold...and then with age they magically melt into elegant, velvety smooth, deeply complex creatures which hint at flavors of ripe fruit, tea, tar, and leather. Wow.

This is one of the few wines that I have encountered that displays its fruit component much better after a decade than it does ten minutes off the vine. And it can be a magnificent adventure too. Like its Bordeaux and Burgundy counterparts, this wine is not crafted for quick consumption. Time must be endured in order for the wine to achieve its best expression, and then when the time comes, take all the time you can to enjoy them.

Nebbiolo is actually quite easy to understand too. While it is ultimately complex on the palate, it is infinitely simple to learn the facts. Unlike Sangiovese, clonal variation is not that big a deal. Unlike other Italian or even French wines, there is not a lot to interpret on the label. And unlike Cabernet Sauvignon and Pinot Noir et al, it is really only produced in this one region. It truly is the grape/wine that has the most geographic expression of them all: it's one distinct grape, making one distinct wine, from one distinct place: how much more distinct can you get?

Perhaps knowing just a few more quaint facts about the grape will help you understand it even more. Nebbiolo derives its name from *nebia*, the Italian word for *fog*. Like in San Francisco, fog regularly envelopes the foothills of the Piedmont region during the harvest season, thus enshrouding the vineyards as well. When you travel the roads in the northwestern part of Italy, and along the western coast in general, you will often see triangle-shaped warning signs that say 'Nebia!'—and now you know what to look for! And you will always think of the wine when you see them, just as I do. In fact, it's the only arcane Italian word I know, just because I saw the signs, and they made me think of the wines. Or was it because I saw the signs while in the car driving while drinking the wines... hmmmm... I can't really remember, but thankfully there was no fog that day.

Anyway, this fog creates an extremely unique growing environment (or *terroir*) in the Piedmont which leads us to the last interesting fact: it's a hard environment to duplicate. So hard in fact, that Nebbiolo is almost the exclusive property of this region, meaning you won't see them produced hardly anywhere else. Given the popularity and prices of these wines, it's not surprising that others have tried, but no one but the Italians pull it off in any volume. It's still a Piedmont thing.

So little production; so geographically limited; so expensive. So why does Boyer want you to know this grape? You have to know the King! Everyone must know the King! And when your palate and your wallet are ready for the King, you too will bow down before the King! It's good to be da' King!

Distinct Nebbiolo descriptors: Blackberry, licorice, cherry, truffle, tar, tobacco, violets, earthy,

General Nebbiolo descriptors: Almond, blackberry, cherry, floral, fruitcake, licorice, perfumed, plum, potpourri, raspberry, roses, rose petal, spicy, tea, tar, tobacco, truffle, violets Woody/bottle age descriptors include oak, smoke, toast, tar, vanilla, earth, leather, cedar, cigar box

Where Nebbiolo is grown: As suggested above: the Piedmont region of Italy is the almost exclusive realm of the King. You will find very limited offerings of this grape from far-flung places like Argentina, Australia, California, New Zealand, and interestingly enough Virginia. I swear that I'm not just promoting the local industry, but check out Barboursville Winery outside Charlottesville for what is easily the best representation of this grape on our continent. I guess it has something to do with the fact that Barboursville is owned and operated by a group of Italians. Que bella!

What foods are often paired with Nebbiolo: Strong cheeses: the heavy hitting kind. Hearty pasta dishes. Deep red sauce dishes. Any beef dish, steak prepared just about any way you want…but the more meaty flavor to come through, the better. Duck, goose, wild game, especially wild boar. The richest, strongest-flavored meats and stews will all bow down before the King.

And let's do some specific cheeses too
Definitely do: Fontina, Grana Padano, Parmesan Reggiano, Fiore sardo
Will work: Gorgonzola, Castelbelbo, Castelmagno, Fontina, Taleggio
Possibly try: Stilton, Aged Cheddar, Aged Gouda, XX Aged Gouda, Smoked Gouda, Chevrelait

Barbera
Not royalty, but still a definitively Italian expression, Barbera is second only to Sangiovese as a workhorse grape of the entire peninsula. It's vigorous, reliable, disease-resistant—let's face it, it's a mule. Even though Nebbiolo gets all the attention in the Piedmont, Barbera accounts for fifteen times as much acreage; almost half of all the vines of the entire region! An extremely versatile and hardy grape, Barbera has mostly been used for

bulk wines and blending.   It always shows up at harvest with intense fruit flavors and enough serious acidity to make it a champion blending option—it can beef up any wine lacking in any attribute.

Unlike Nebbiolo, Barbera adepts well to lots of situations, and as such used to be a major player in the California wine industry of the past, and it's making a comeback right now. Most California 'jug wines' have Barbera in them; some have trace amounts, some have a whole heck of a lot more—but you won't often see the grape name appear anywhere on the label of those bulk wines. But don't let me downplay them. Barbera has been made as varietal wines in the past, and they are rapidly gaining ground as high-end expressions of the region….and further abroad too.

Back in Italy, and increasingly in California, the Barbera is being crafted into some very interesting expressions. The best of the bunch still come from the Piedmont: look for labels that say *Barbera d'Asti, Barbera del Monferrato,* and *Barbera d'Alba.* Can you now guess what these labels mean? *Del, di, de,* and *D'* simply mean of; so these are Barbera of Asti, Barbera of Montferrato, and Barbera of Alba—all towns in the Piedmont region, well worthy of visitation. Great food, great wine, great people. And now you know how to decipher those labels too. Party on! Barbera does produce a lot of color (dark ruby reds), a lot of acid, and a lot of rich, fruity, berry flavor, and therefore may not appeal to all. It is typically crafted for consumption when young, and I must be honest when I tell you that I don't personally like those over-ripe, jammy flavors in a wine, but many do. A lot of them are tart and light, and can even be a little fizzy due to their youthful and energetic fermenting exuberance. If you get one that taste like that, it is probably very young, has not seen any oak, and that's the way it's supposed to be. But when aged for a bit in oak, by the careful hand of a good winemaker, these grapes can produce some complex aromas and tastes that bring out those bright flavors with a smoothness that is divine…and usually for a good price!

Distinct Barbera descriptors: Blackberry, raspberry, cherry, red berries, juicy, jammy (especially when young).

General Barbera descriptors: Blackcurrant, blackberry, raspberry, hackberry, all red berries, gooseberry, plum, violets, tart, acidic. Woody descriptors: vanilla, toasty, oaky

Where Barbera is grown: Again, the last of the stallions is a Piedmont player—but it can be found all throughout Italy: Lombardy, Emilia-Romagna, and Sardinia in particular. Argentina seems to be producing a lot more of them as well. As referenced above, Barbera used to account for large acreages in California, and some of the best varietal expressions are now coming out of the Central Valley, Sonoma, Napa, the Sierra Foothills, Paso Robles, Santa Clara and Sonoma.

What foods are often paired with Barbera: Mild to strong cheeses, seafood in heavier cream sauces, particularly if they are red or brown. Salmon, tuna steaks, game bird, heartier pastas, beef dishes, barbecues, duck, goose, grilled pork, Italian sausage, pork roast, ham, cold cuts, a solid good pizza. I like a well-crafted Barbera with a hearty meat lasagna.

And let's do some specific cheeses too
Definitely do: Abbaye de Belloc, Banon, Fontina
Will work: Fiore Sardo, Grana Padano, Lancashire
Possibly try: Ossau-Iraty, Piave, Taleggio Raclette, Feta, Gorgonzola

<u>**Additional barrel flavors**</u> you may detect in any of the wines depending upon how much time it spent in oak: (From Lesson 4: Got Wood? The Influence of Oak)
  ➢ *Specific to these red wines*: Almond, bacon, burnt, cashew, cedar, coconut, nutty, pencil shavings, sawdust, smoky, spicy, toast, toffee, vanilla
  ➢ Woody descriptors: Cedar, charred wood, cigar box, coffee, green wood, oaky, pencil shavings, resinous, sandalwood, sawdust, smoky, toasty, tree bark, vanilla.
  ➢ Barrel flavor (malolactic fermentation) descriptors: Almond, banana, burnt caramel, butter, buttered popcorn, butterscotch, cashew, coconut, lanolin, vegemite
  ➢ Lees flavor descriptors: Baked bread, creamy, cheesy, leesy, nutty, yeasty, yogurt

Some final thoughts on these wines:
Try to focus on the fruit components as you do this exercise, knowing full well that everything else changes according to region, climate, and winemaker influence. But with these wines, regional variation is not as key since so many of them are exclusively Italian affairs.

Let's recap a bit on the other stuff:

Clones are a slight genetic variation of a grape variety. This actually happens to all grape varieties (you might see reference to Chardonnay Clone 6: a popular child in its category) but only in a few like Sangiovese do you see such attention paid to the clones that they would get their own name. Sangiovese has at least 14 distinct clones, many of them named. Which brings us to names:
  • **Sangiovese synonyms** to remember: Chianti, Chianti Classico, Chianti Superiore, Morellino di Scansano, Vino Nobile de Montepulciano, Brunello di Montalcino. (sub-regions of Tuscany). Most are 100% Sangiovese clones, while some are blends using 80-90% Sangiovese.
  • **Nebbiolo synonyms** to remember: Barolo and Barbaresco (sub-regions of Piedmont). Barolo is deep and rich, while Barbaresco is usually light and more accessible. Both are 100% Nebbiolo.
  • **Barbera synonyms** to remember : Barbera d'Asti, Barbera del Monferrato, and Barbera d'Alba. Named for the towns of their

production/near their production.

And one other term that I haven't introduced you to yet, but that you certainly want to know because it's a hot catch-phrase: **Super-Tuscans!** What the hell is a **Super-Tuscan?** Italy, like France, has strict laws based on the geographic zones (i.e. Chianti) of production which dictate the types of grapes can be used, and how much of each type of grape, as well as a host of other things. Back in the late 80's and 90's, many high-end winemakers in Tuscany got tired of the confines of this system, and decided to forgo their regional labels to be able to free to do what they wanted, and to experiment. So they started crafting very kick-ass wines by blending Sangiovese with Cabernet Sauvignon and Merlot, among other experiments. And man, are these things good! Even though they may officially be located inside the Chianti Classico sub-region, they weren't using the specified grape formula, and therefore were not allowed to put the region name on the label.

However, the guys who did this were already established producers, and their names and reputations allowed them to demand high prices for these 'super' wines—prices which they promptly got! Because the wines rocked! And thus was born the term 'Super-Tuscan': a great wine, from a great producer, in a classic region, who is simply not following the rules of the region! These guys were so successful at it, that many of the laws have been relaxed so as to now allow them to label regionally again. You will often hear reference to this term, and know this: if you see an Italian wine at a store that is very expensive, but has no regional denomination, then it's probably one of these Super-Tuscans. You may even hear reference to some California winemaker's mimicking a super-tuscan style!

When thinking about the big three grapes from Italy, we can make some French comparisons. Remember the Italian Stallions like this:
  - ➤ Nebbiolo is to the Piedmont what Pinot Noir is to Burgundy: 100% of the variety, labeled by the region.
  - ➤ Sangiovese is to Tuscany as Cabernet Sauvignon is to Bordeaux: the main component of the blend that is labeled for the region.
  - ➤ Barbera is to the Piedmont/California as Merlot is to Bordeaux: a workhorse, reliable grape that is used to blend with other greater grapes (but, of course, sometimes stands on its own).

Got it? Then Party on!

	Comments
Sangiovese	
Color:	
Bouquet (smell):	
Taste:	
Body:	
Compare/Contrast to other wines:	
Nebbiolo	
Color:	
Bouquet (smell):	
Taste:	
Body:	
Compare/Contast to other wines:	
Barbera	
Color	
Bouquet (smell):	
Taste:	
Body:	
Compare/Contrast to other wines:	

Lesson 14: The Spanish Princes: Grenache & Tempranillo

What they smell like. What they taste like. Where you get them from. What foods bring out the best in them. And many of the aliases that they go by.

Buenos días amigos! Today we will be intimately investigating into Iberia for this round of varietal tasting, and I don't know if any of the lessons you have done so far have consisted of such polar opposites as these two players in the Spanish grape pantheon. Grenache and Tempranillo are about as far removed from each other as grapes can get in terms of aromas, tastes, tannins, body, color, age-worthiness, and even global distribution. But as with all things in life, sometimes opposites attract! And these two grapes are often harmonized together in a lot of specific regional wines for which Spain is famous.

In this tale of brotherly love, Tempranillo is definitely the suave, handsome, classy older brother to Grenache's grungy, back-ally-brawling, harder-working but lesser respected, youthful junior position. Tempranillo will definitely be ascending to the throne to be king, while Grenache lurks around the back rooms of the palace…but he is always there! As with all such relationships, Grenache is much better known and traveled in the world…he is a party animal…while Tempranillo has mostly been confined to the castle as an aristocrat, and is only now becoming known outside the royal domain…

But that's what this tasting is all about! Let's get to know these guys! Are we ready? Then vamanos! Vamanos! Vamanos!

What to grab:
- One bottle from each list of the following from your local wine shop or grocer (or just do one at a time at your own pace…because these guys can get expensive, fast):
  Grenache
  Bottles $5-$15
  - ➤ Deisen (Australia)
  - ➤ Borsao Tinto (Spain)
  - ➤ Tres Ojos Old Vines 2005 (Spain)
  - ➤ Las Rocas (Spain)
  - ➤ Sella & Mosca Cannonau di Sardegna Riserva (Italy)
  Bottles$16-$30
  - ➤ Chalone Vineyard Gavilan (California)
  - ➤ Kilikanoon Prodigal Grenache (Australia)
  - ➤ Tour De L'Isle Gigondas (70%) (France)
  Bottles $30+
  - ➤ Santa Duc Gigondas (80%) (France)
  - ➤ Alto Moncayo Moncayo (Spain)
  Tempranillo
  Bottles $5-$15

> ➤ Finca Sobreno Toro (Spain)
> ➤ Marco Real (Spain)
> ➤ El Arte De Vivir (Spain)
> ➤ El Circulo Crianza (Spain)
> ➤ Berberana "Dragon" de Castilla (Spain)
> ➤ Campo Viejo Reserva Rioja (Spain)

Bottles$16-$30
> ➤ Valsacro Cosecha Rioja (Spain)
> ➤ Bodegas Vizcarra Ramos J.C. (Spain)
> ➤ Prima Toro (Spain)
> ➤ Erial Ribera Del Duero (Spain)

Bottles $30+
> ➤ Erial TF 2005 (Spain) Wow. One of my favorite wines ever.
> ➤ Villacreces Nebro (Spain)

Don't worry if you can't find these exact labels. You can always ask your local wine store salesperson to help you pick out a good representation of the varieties I have described above. Show them the list and tell them to get you something close to it.

What to do:
1. First open the bottle(s) and try immediately. What is the color, the aroma, the body, and the taste of each wine? Record all impressions. Write down some descriptions of your experience, both what you smell and how it tastes. You should be able to detect some subtle difference in flavors and aromas in these wines…and perhaps even different colors as well. Really work hard at picking out and describing differences between each of these varietal wines. As always, include as many descriptors as possible to define each wine, even if some may be a stretch. Check below for a bit of varietal vocabulary to choose from.
2. Next wait at least 2 hours for the wine to "open up," leaving the cork off the wine to allow oxygen to get in. Try the wine again. Has the smell changed? How so? What about the taste? Is it fruitier? Does it taste better to you? As you go back and forth from one wine to the next a few times, be sure to cleanse your palate with a bit of bread or plain crackers. Sip some water as well, if you absolutely must. But don't eat any other food yet—it will easily alter the results of your descriptions. Don't worry: you can finish the bottles with dinner later.
3. Now let's revisit Lesson #1: try the wines with cheese or any appropriate food.   Record your reactions. Has anything changed in the taste? Does the wine seem smoother or lighter or fruitier? Repeat with a different type of food if available. Again, detect some differences and make yourself write down some different descriptors.
4. Finally (if possible), save a bit of wine in the bottle. Let it sit exposed (no cork) on the counter for a few more hours. Let it go overnight if you want. Then try it again on its own, without food. What's it like

now? Write it down. Compare this
with your first impression of the
wine right after you opened it. What
has happened?

5. Keep the rest of the bottles for
    dinner. Try with a variety of
    different foods. What works? What
    doesn't? What foods bring out more
    flavors in the wine? What foods
    overwhelm the wine? I want you

to experiment as much as you want to see what works best for you,
but listed on the next several pages are some variety/food pairings
that do seem to bring out the best in these particular reds that you
might want to try.

What to look for:

The whole point of this exercise, and others in this section, is to get you
fluid and fluent in recognizing and describing the major varietal wines
produced in the world today. Of course you know that the term *varietal
wine* means: a wine made predominately, if not exclusively, from one type
of grape. As you do the tasting, focus on comparing and contrasting these
different wines, with more of an emphasis on contrasting.

Later exercises in this Boyer tasting round-up will compare some varietal
wines from different regions, but you should do them on your own as

often as possible too. When purchasing a bottle of one of these wines for dinner, go ahead and grab an inexpensive version of the same grape from a different part of the world. Your knowledge will grow exponentially, even if it is at the expense of your wallet decreasing mathematically. This is easy to do with Grenache since it is one of the most widely planted grape varieties on the planet. However, it will be much more challenging for you to find a Tempranillo outside of Spain. Much like Nebbiolo is to Italy, Tempranillo is a distinctly Spanish affair…and once you have a really good one, you will want to run off and elope with it too!

The Specifics:

Let's take a look at each of these major Spanish varieties one at a time to provide some tasting and smelling clues, some regional background, and also some food paring hints. Let's start with…

Grenache

Let's start with the grungy underling first. Why do I mock Grenache so? It's all friendly banter here, but this is one of those grape varieties that simply will never make it to the upper classes of wine aristocracy. It's a peasant really; a tireless workhorse grape that may possibly be the most planted grape in the world. You can find tens of thousands of acres across Spain (where it is the most planted grape variety), in southern France, in California, South Africa, and even in Australia. But you've probably never heard of it, and less likely than that to have seen its name on a wine label. Why is that?

Because it is generally considered a lesser-quality varietal wine; to be honest, Grenache has got several strikes against it. While it grows like a weed pretty much anywhere it's thrown down in warm or hot climates, the resulting fruit and juice of Grenache is naturally low in tannins and color and even malic acid…however, it will produce fruity flavors and sugars like a champ! So the end product is a wine that is typically light colored, way high in alcohol, with sweet berry characteristics that have a spicy edge. Well hold the phone! High alcohol and fruit flavors and spice…give me a double! What's wrong with that?

Just this: if you remember your lesson on balance and body, you might be tipped off already that this high alcohol/low acid/low tannin combination doesn't usually equate to a fantastic finished product. Also, because of the lack of tannins and acids in particular, Grenache wines are prone to oxidation quickly, and they age rapidly. Their fruit flavors will also disappear rapidly over time. They've got no legs to go the distance! So they are best consumed young. But that's okay, because that's when they are at their best…in their heady youth, Grenache wines are hot with alcohol, fleshy, fruity, plumy, and spicy-licious! Some people really dig them.

But most people really dig them without even knowing. Because the biggest use of Grenache is as a component of blended wines. That high alcohol and deep red fruit flavor has been employed by winemakers for centuries 'fill out' or 'beef up' red blends and particularly to soften big tannic wines like Syrah. In Spain, it's blended in some amounts into a vast array of different wines from different regions. In France, Grenache makes its way into many of the wines of Provence, Languedoc, and the southern Rhône regions—and even is a part of the most famous wines of the Rhône, the Chateneuf-du-Pape. Grenache can't be all bad if it's in du Pape! And oh how I love the Pappy!

Grenache's sweet berry flavors also make it an ideal candidate to make some very pleasing rosés—you may have had some on previous lessons already. In southern France, look for pink wines labeled Lirac or Tavel (regional names) which are fantastic because the sweet berry flavors are maintained in the largely dried out wine, making for a very nice lightly fruity/spicy but not sweet product.

So don't look for greatness from this varietal wine, but what it lacks in character, it more than makes up for in utility. Go ahead and try one or two so that you see what it is like on its own…and you will be able to pick up on its attributes in other blended wines as you progress forward on your wine drinking path!

Distinct Grenache descriptors: fruity fruity fruity + spicy spicy spicy! Jammy, pepper, red current, raspberry, fleshy, rustic, sweet berry.

General Grenache descriptors:
Blackberry, black cherries, black pepper, licorice, meaty, pepper, plum, plumy, prune pruney, stewed prunes, raspberry, spicy, tar, tobacco, barnyard, earthy, gamey.

Where Grenache is grown: As suggested above: everywhere! The grape probably has its origins in Spain, and makes it way into wines from virtually all of Spain's winegrowing regions, but especially Rioja…and look for exceptionally well-crafted 100% Grenache wines from a small region named Priorat. It is also prolific across southern France. Look especially for Rhône Valley examples, and the rosés from Tavel and Lirac. Still tons of it in California, although much less than there was 30 years ago when it used to be a staple of the bulk red wine industry. Same goes for Australia. Also some interesting versions from Sardinia

What foods are often paired with Grenache: Mild to strong to very strong cheeses, garlic rubbed spareribs, chicken livers with caramelized onions and Madeira, Korean barbecued pork, Grilled lamb with radicchio and black-olive oil. Wow! BBQ chicken wings, pancetta, prosciutto, salami, mild Italian sausage. Soups like bean or lentil, French Onion, gazpacho, pasta e fagioli. Seafood: shrimp, scallops or calamari in a tomato based sauce, mahi mahi. Chicken with mushrooms, roasted game hen, pheasant.

Cheeseburger, chesesteak, steak calzone.

And let's do some specific cheeses too
Definitely do: Azeitao, Ossau-Iraty, Jarlsberg
Will work: Tete des Moines, Cheddar (Mild), Gouda, Swiss
Possibly try: Fontina, Habenero Jack, Edam

Other names for Grenache: Garnacha or Garnacha Tinta in Spain;
Grenache Noir, Garnacha, Cannonau, Lladoner, Tinto Aragones, Alicante

Tempranillo

Okay...we let the peasant wash the floors and help build the body of some
great Spanish and French red blends...but now onto the true blueblood!
Tempranillo is the backbone of some of the greatest Spanish wines from
the Rioja, Ribera del Duoro, Cataluna, and Toro. And it is increasingly
making headway into the all-star world line-up as a stand-alone varietal
wine. This stuff totally rocks! Tempranillo is Spain's answer to Cabernet
Sauvignon in France and Nebbiolo in the Piedmont of Italy: it is the king of
its prospective wine country!

And you want to know the best part of it? The world is just now figuring
this out! So while Tempranillo is still largely produced in Spain,
experimentation and planting have just begun in other parts of the world.
Which means that new and exciting versions of this varietal wine will
be popping up around the planet for some time to come. But for now
you will just have to get the great examples of this wine from its Spanish
homeland...and then you get to look really smart by impressing all your
wine-drinking friends by pulling one of these bad boys out at a dinner
party. Because most folks haven't even had one yet!

Tempranillo seems to at once capture the delicate brightness of a Pinot
Noir and the fleshy fruitiness of a nice Merlot. The grape has a great
balance of an acidic/tannic backbone with unique depth of subtle, scented
fruit flavor. The body tends toward the rich and velvety side after some
barrel aging, which is typical for varietal versions. This combination of
characteristics make these wines particularly age-worthy as well, and boy
oh boy, they continue to get smoother and more subtle and complex with
the years. Yeah!

But those 100% pure varietal versions are actually about as common as
the Grenache ones are. Like its little brother, Tempranillo is most often a
component of blended reds. Unlike its little brother, it is usually present
in the greatest concentrations, as opposed to just being a supplement.
In fact, it's a Tempranillo/Grenache combo that makes for some of the
greatest wines of Rioja. Especially when barrel aged.   In fact, I should warn
you: Tempranillo is fine on its own, but a lot of its great characteristics
only come together after some integration in the oak barrel. Oak and
Tempranillo just work magic together.

And watch out my friends! Folks are starting to experiment even more with blending options for Tempranillo too. I've had some Tempranillo/Cabernet Sauvignon blends that knocked my socks off! Talk about a pair of kings as the game-winning hand! I'm always ready to ante up if a Tempranillo is at the table!

Distinct Tempranillo descriptors: Dark cherry, herbal, plum, tobacco leaves, spice, violets, earthy.

General Tempranillo descriptors: Dark cherry, dark berry, dried strawberry, blackberry, cherry, floral, licorice, perfumed, plum, raspberry, spicy, tea, tar, tobacco, truffle, violets, berryish fruit, herbaceousness, and an earthy-leathery minerality. Woody/bottle age descriptors include tobacco leaf, oaky, vanilla, earth, leather, nutty, cocomut, sandalwood.

Where Tempranillo is grown: All regions of Spain, but especially known in Rioja, Ribera del Duoro, and Cataluna. Is also widespread across Portugal, where it goes under a variety of names and is used in Port wines. Big plantings are also found in Argentina. Some small plantings in southern France. Not too much in the New World though.

What foods are often paired with Tempranillo: Lamb, roasted duck, beef stew, small plates of Chorizo, olives, shrimp Tortilla Espanola, Confit duck leg, spiced red cabbage, cotechino sausage, watercress and saba dressing, tortilla soup, lamb meatballs with cumin, mint, and tomato sauce… Hearty pasta dishes. Deep red sauce dishes. Any beef dish, steak prepared just about any way you want…but the more meaty flavor to come through, the better.

And let's do some specific cheeses too
Definitely do: Roncal, Garrotxa, Idiazabal Cheddar, Havarti
Will work: Montgomery's Cheddar, Manchego, Grafton Village Cheddar,
Possibly try: Cheddar (sharp), Manchego, Zamarano

Other names for Tempranillo: Cencibel in La Mancha; Ull de Llebre in Catalonia; Tinto Fino in the Zamora region; Tinta del Pais in the Ribero del Duero; Tinta de Toro in the Toro region. In Portugal, Tinta Roriz to port producers and Aragonez elsewhere.

Some final thoughts on these wines:
Try to focus on the fruit components as you do this exercise, knowing full well that everything else changes according to region, climate, and winemaker influence. But with these wines, regional variation is not as key since so many of them are exclusively Spanish or southern France. You shouldn't have any trouble telling them apart in terms of different tastes… it is night and day.

To recap a bit:
These two grape varieties are more often than not components of

blended red wines in Spain and France, although varietal expressions are increasing.

If you like Grenache flavors and aromas, you are sure to find them in trace amounts in almost all Spanish and southern France wines…including some big hitters like Chateneuf-du-Pape. Or look for a regional wine called Gigondas, which is typically 80% Grenache and almost as totally awesome as the du-Pappy.

You can easily find some great Grenaches in the form of rosés from the southern Rhône regions of Tavel and Lirac.

Tempranillos are typically much much better after some oak aging, as are most of the big-hitting Spanish reds.

**Additional barrel flavors** you may detect in these wines depending upon how much time it spent in oak: (From Lesson 4: Got Wood?)
   ➢ *Specific to these red wines*: Almond, bacon, burnt, cashew, cedar, coconut, nutty, pencil shavings, sawdust, smoky, spicy, toast, toffee, vanilla.
   ➢ Woody descriptors: Cedar, charred wood, cigar box, coffee, green wood, oaky, pencil shavings, resinous, sandalwood, sawdust, smoky, toasty, tree bark, vanilla.
   ➢ Barrel flavor (malolactic fermentation) descriptors: Almond, banana, burnt caramel, butter, buttered popcorn, butterscotch, cashew, coconut, lanolin.

Let's add just a bit more about wood here too:
I have already told you that Tempranillo in particular benefits greatly from time in oak, and so let me give you just a few terms you will see on Spanish wine labels which alert you to the oaked status of the contents. All these wines may be the exact same blend of Tempranillo and Grenache, from the same vintage, and maybe even the same vineyards, but:
   ❖ A wine from Rioja that has the word *joven* (young) on the label indicates a wine that has not seen any time in oak and was releases less than a year after its production
   ❖ A wine from Rioja that has the word *crianza* on the label indicates a wine that, by law, has been aged for 2 years with at least one of those years in oak barrels.
   ❖ A wine from Rioja that has the word *reserva* on the label indicates that , by law, has been aged for 3 years with at least one of those years in oak barrels.
   ❖ A wine from Rioja that has the word *gran reserva* on the label indicates that , by law, has been aged for 5-7 years with at least two of those years in oak barrel.

I'll let you figure out how those designations affect the prices of these wines. And they really have to tell you this stuff, because the addition of barrel aging really makes for a radically different product of these Spanish princes. Do another experiment on those some time on your own to see for yourself. And I don't want to confuse you, but the age descriptions listed above are just for Rioja wines. There is no set standard for age labeling across Spain…or really anywhere else now that I think about it… but a lot of producers across Iberia are starting to use this basic labeling convention that I believe you will increasingly see as Spanish wines continue to be hot on the international drinking scene:

Noble indicates the wine in the bottle spent 1 year in oak.

Anejo indicates the wine in the bottle spent 2 years in oak.

Viejo indicates the wine in the bottle spent 3 years in oak.

Got it? Then get busy drinking those aged Spanish wines! And…Party on me amigos!

	Comments
Grenache red	
Color:	
Bouquet (smell):	
Taste:	
Body:	
Compare/Contrast to other wines:	
Grenache rosé	
Color:	
Bouquet (smell):	
Taste:	
Body:	
Compare/Contast to other wines:	
Tempranillo	
Color	
Bouquet (smell):	
Taste:	
Body:	
Compare/Contrast to other wines:	

Lesson 15: The Rhône Rangers: Syrah & Mourvèdre

What they smell like. What they taste like. Where you get them from. What foods bring out the best in them. And many of the aliases that they go by.

No time for Tonto….let's just talk about the Rhône Rangers! Why the Rhône Rangers? Well, these two grapes are primary players in almost all the Rhône region wines…but their stewardship ranges much further than just southern France! The workhouse Grenache, along with our new friends Syrah and Mourvèdre, are widely planted across all of Spain and southern France and form the backbone of a famous blended wine that you will see referred to in various regions as **GSM**. See if you can figure out what that acronym stands for over a bottle of wine sometime. But the Iberian Peninsula and vicinity are only the start of their world adventures…

In fact, you need only consider Syrah for a moment. Perhaps you've never even heard of Syrah yet…but how about Shiraz? Of course you have! That's the Australian stuff! But it's exactly the same grape! That's right: Syrah = Shiraz. And while we are at it, Old World Mourvèdre = Mataro in the New World. But not just the names changed as our rangers headed abroad. These grapes have traveled the world over, and become something radically different stylistically in their journeys. Australian Shiraz wines are a million miles away from Rhône Syrah wines in aromas and flavors and bodies: a difference you are about to find out about right now if you so choose (if not now, then in Lesson 17 you will).

So let's try to catch up with these two varietal rangers and find out what they are all about. Are we ready? Then: Hi-ho silver….and away! Let's drink this now!

What to grab:
- One bottle from each list of the following from your local wine shop or grocer…with the exception that you really only need to do one version of Syrah/Shiraz even though there are selections on two different lists. I just wanted you to be aware that even though it is the same grape, the stylistic difference between Old World and New World is huge. If you want to go for the gusto, go ahead and get one of each, along with the Mourvèdre, and check it out for yourself.
  Syrah (Old World-style: Europe)
    Bottles $5-$15
    - Chateau Grande Cassagne Hippolyte (France)
    - Puig Parahy (France)
    Bottles $16-$30
    - Qupe (California)
    - Chateau Guiot Costieres de Nimes (France)
    - La Reverence (France)
    Bottles $30+
    - Barrel 27 (California)
    - L'Ecole Nº 41 Seven Hills Walla Walla Valley (Washington)

Shiraz (New World-style: Australia)
Bottles $5-$15
- Milton Park (Australia)
- Step Road Blackwing (Australia)
- Rosemount (Australia)
- Peter Lehmann (Australia)

Bottles $16-$30
- Thorn Clark Shotfire (Australia)
- Torbreck Woodcutters (Australia)
- Mollydooker The Boxer (Australia)

Bottles $30+
- Two Hands Lily's Garden (Australia)
- D'Arenberg Dead Arm (Australia)
- Two Hands Ares (Australia)

Mourvèdre
Bottles $5-$15
- Jade Mountain (California)
- Castano Monastrell (Spain)
- Bodegas Castano Hecula (Spain)

Bottles $16-$30
- Finca Luzón Jumilla Altos de Luzón (Spain)
- Cline Ancient Vines (California)

Bottles $30+
- Hewitson Old Garden (Australia)
- Bonny Doon Old Telegram (California)
- Domaine Mathieu Chateauneuf du Pape Vin Di Felibre 2003 (France) (this one is 80% Mourvèdre, which is a very high % for a Rhône wine)
- Mas De Boislauzon Chateauneuf Du Pape Le Tintot (France)

Don't worry if you can't find these exact labels. You can always ask your local wine store salesperson to help you pick out a good representation of the varieties I have described above. Show them the list and tell them to get you something close to it.

What to do:

1. First open the bottle(s) and try immediately. What is the color, the aroma, the body, and the taste of each wine? Record all impressions. Write down some descriptions of your experience, both what you smell and how it tastes. You should be able to detect some subtle difference in flavors and aromas in these wines…and perhaps even different colors as well. Really work hard at picking out and describing differences between each of these varietal wines. Include as many descriptors as possible to define each wine, even if some may be a stretch. Check below for a bit of varietal vocabulary to choose from.

2. Next wait at least 2 hours for the wine to "open up," leaving the cork off the wine to allow oxygen to get in. Try the wines again. Again, really work hard at picking out and describing differences between

each of these varietal wines. Especially if you choose to do both the Old World and New World versions of the Syrah. As always, include as many descriptors as possible to define each wine, even if some may be a stretch. As you go back and forth from one wine to the next a few times, be sure to cleanse your palate with a bit of bread or plain crackers. Sip some water as well, if you absolutely must. But don't eat any other food yet—it will easily alter the results of your descriptions. Don't worry: you can finish the bottles with dinner later.

3. Now let's revisit Lesson #1: try the wines with cheese or any appropriate food. Record your reactions. Has anything changed in the taste? Does the wine seem smoother or lighter or fruitier? Repeat with a different type of food if available. Again, detect some differences and make yourself write down some different descriptors.

4. Finally (if possible), save a bit of wine in the bottle. Let it sit exposed (no cork) on the counter for a few more hours. Let it go overnight if you want. Then try it again on its own, without food. What's it like now? Write it down. Compare this with your first impression of the wine right after you opened it. What has happened?

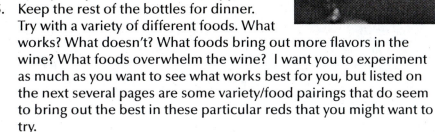

5. Keep the rest of the bottles for dinner. Try with a variety of different foods. What works? What doesn't? What foods bring out more flavors in the wine? What foods overwhelm the wine? I want you to experiment as much as you want to see what works best for you, but listed on the next several pages are some variety/food pairings that do seem to bring out the best in these particular reds that you might want to try.

What to look for:

The whole point of this exercise, and others in this section, is to get you fluid and fluent in recognizing and describing the major varietal wines produced in the world today. Of course you know that the term *varietal wine* means: a wine made predominately, if not exclusively, from one type of grape. As you do the tasting, focus on comparing and contrasting these different wines, with more of an emphasis on contrasting. But know this: these particular varieties are also widely used in lots of regions to produce **blended wines** which are of course wines not made of a single grape variety, but a blend of two or more grape varieties. The **GSM** mentioned in the intro is an example of a blended wine.

Later exercises in this Boyer tasting round-up will compare some varietal wines from different regions, but you should do them on your own as often as possible too. When purchasing a bottle of one of these wines for dinner, go ahead and grab an inexpensive version of the same grape from a different part of the world. Your knowledge will grow exponentially, even if it is at the expense of your wallet decreasing mathematically. This is easy to do with Syrah in particular, since so many folks are growing it, and making so many different styles out of it as well.

The Specifics:
Let's take a look at each of these major Spanish varieties one at a time to provide some tasting and smelling clues, some regional background, and also some food paring hints. Let's start with…

Syrah
Que-syrah, syrah! Whatever will be, will be! And this grape's future is quite secure for whatever will be the future of the wine world. Syrah forms robust, intense wines of a variety of styles. Almost always deep violet to nearly black in color with chewy texture and richness, Syrah wines are not for the faint of heart. Their often high-octane alcoholic strength and high tannic component also make them daunting when young…but, of course, extremely age-worthy. Over time these wines can evolve into magnificence. But with many styles, especially the Australian versions, you don't even have to wait! Drink them now, and enjoy the fruity and spicy aromas and flavors that have been smoothed out with oaky-vanilla overtones. But I'm getting ahead of myself….

Syrah has long been considered one of the original 'noble' grapes of Europe; a variety with a recognized historical pedigree. Syrah is the grape that makes for some of the most famous—and expensive—wines in the northern Rhône region, namely *Hermitage*, Côte Rôtie, and Cornas. As mentioned previously, Syrah is also the foundation block of the GSM blend which is the backbone for most Rhône blends, including the ever popular **Chateauneuf du Pape**. But to be honest, not many outside of France (except the rich and famous) had even sampled a Syrah wine until well into the 20th century.

Enter: those whacky Australians! Adopting and renaming the grape with their own peculiar contrived accent, *Shiraz* was born, Australian-style! The incredible popularity and success in the 'land down under' is what made Syrah into the international success story it is today. See, many decades ago the Australians committed themselves into becoming a major global exporter of wines; a goal they have largely achieved. And they have achieved this mostly by making Shiraz a highly drinkable, highly marketable, and highly identifiable varietal wine. Kudos to you Aussies! Throw a shrimp on the barbie to celebrate!

However, the Australian Shiraz success story only came about because they invented their own style for the Syrah grape. The Frenchies always made their wines with more accent on the tannin and spice, making it a wine which needed to be aged a bit in order to mellow out the flavors. But the Australians wanted to craft a wine that was ready to go right this second, so their wines are what we call fruit forward. They also integrated much more of the vanilla, oaky flavors from the barrel which make for a creamy, more approachable wine on the spot.

Quick Syrah summary:
Old World Syrah: accentuate the spicy over fruity; very long-lived, tannic and spicy red wines which typically need some age to integrate and soften its components.
New World Shiraz: accentuate the fruity over spicy; much more intensely fruit driven with less hard-edged, sharp tannins and spice...even in youth.

But whether New World or Old, Syrah produces rich, dark, complex, intense, full-bodied reds which almost have to be chewed before swallowing! And even though many of those Australian versions are ready to drink right now, all well-crafted Syrah/Shiraz have enough color, tannin, and alcohol to go the distance in a cellar as well. So enjoy them now, enjoy them later, or enjoy them a decade from now. Wow! This Rhône Ranger is unstoppable, over space and time!

Distinct Syrah descriptors: Black current, blackberry, grass, black pepper, licorice, clove, thyme, bay leaf, sandalwood, cedar, musk, civet, earthy, vanilla, coconut, sweetwood, cigar box, leather, tar, toast, oak.

General Syrah descriptors:
Aniseed, blackberry, black cherries, black olive, black current, black pepper, blueberry, cinnamon, clove, eucalyptus, herbs, licorice, menthol, mint, mulberry, nutmeg, pepper, plum, plumy, prune, raisin, raspberry, red current, spicy, stewed plums, chocolate, earthy, leather, meaty, mushroom, salami, burnt, cedar, toast, oak, tar, tobacco, barnyard, sawdust, earthy, gamey.

Where Syrah is grown: As suggested above: everywhere! While Australia may grow tons of the stuff, the grape probably has its origins in the Rhône, where it still makes those big boy wines mentioned above, namely **Hermitage**, Côte Rôtie, and Cornas. And France still grows the most, although acreage has been growing fast in California, Southern Oregon, Washington, Spain, Argentina, Chile, and South Africa.

What foods are often paired with Syrah: Mild to strong to very strong cheeses, garlic rubbed spareribs, chicken livers with caramelized onions and Madeira, Korean barbecued pork, Grilled lamb with radicchio and black-olive oil. Wow! BBQ chicken wings, pancetta, prosciutto, salami, mild Italian sausage. Roasted game hen, pheasant. Cheeseburger, chesesteak, steak calzone.

And meat! You can't go wrong with red meat on this one! Duck, goose, game birds, roast lamb, grilled sausage or cassoulet, mushroom pizza, spicy pizzas, herbed sauces on red meat, lamb chops, veal chops, duck breast with plum sauce.

Specifically, steak! Steak! Steak! Rare or well-done, pan-seared or grilled on the open flame! Throw it on the barbie, mate! It is the best with a hearty Shiraz!

And let's do some specific cheeses too
Definitely do: Dubliner, Tomme de Savoie, Reblochon, Beaufort, Roncal
Will work: Saint Nectaire, Cabecou Feuille, Somport, Monbier, Pecorino
Possibly try: Sharp Cheddar, Roaring 40's Blue, Smoked Cheddar

Other names for Syrah: Ummm…..Shiraz. Or have you not picked up on that yet?

Mourvèdre

The Beast is unleashed! This variety is an animal! High in alcohol, high in colors, high in tannins, and high in flavors, this wine can achieve heady… even dizzying…heights of body and flavor which are typically too much for a standard sipper of table wine. And speaking of heady! Mourvèdre can often have a herbal, sage-like, dangerously-gamey-bordering-on-beastly aroma and flavor which sends most running for cover! Ah! I'm scared!

As such, it is primarily used as a blending component in a host of other wines across Spain and southern France. Mourvèdre's bold structure and beastly character add serious 'uumph' to blends, even when present in small quantities. It has the ability to stiffen the resolve of any wine it is blended into.

Along with Syrah, Mourvèdre is an important component in the best examples of the southern Rhône's most famous appellation, Châteauneuf-du-Pape. It also is particularly well suited match to blend with Grenache, softening it and giving it form and structure that it lacks on its own.   And just like its Rhône brother-in-arms Syrah, it has been transplanted to the New World, made into great varietal wines, and renamed. *Mataro* is the American and Australian name for this Iberian brute.

Rough and rustic, deeply colored and distinctly flavored, producing meaty wines with jammy blackberry character…this variety is not one to be trifled with. But if you are up for the battle, add a little animal to your wine diet today!

Distinct Mourvèdre descriptors: Gamey, wild, beastly, earthy, violet, blackberry, sage, thyme, cinnamon, clove, black pepper

General Mourvèdre descriptors: Blackberry, dark cherry, cherries, cinnamon, clove, gamey, green peppercorns, leafy, plum, roses, dark berry, soy, spicy, stewed rhubarb, dried strawberry, violets, chocolate, earthy, gingerbread, leather, toast.

Where Mourvèdre is grown: As with Syrah and Grenache, Mourvèdre is a blending component for various wines across Spain and southern France.

Need some proof? It is Spain's second most planted grape, although you don't often see a whole lot of varietal wines made out of it. California and Australia are also experimenting a lot more with this grape by making a whole bunch of extremely interesting varietal wines right now.

What foods are often paired with Mourvèdre: Pot roast and vegetables, grilled lamb, roasted duck, roasted game, barbecued meat, hearty beef stew, marinated steak with asparagus, hearty venison stew, lamb in apricot sauce, hearty pasta dishes. Deep red sauce dishes, and I mean deep red, like black. Any beef dish, steak prepared just about any way you want… but the more meaty flavor to come through, the better.

And let's do some specific cheeses too
Definitely do: Azeitao, Havarti, Mahon, Zamarono
Will work: Manchego, Chevrotin des Aravis, Serena, Reblochon
Possibly try: Sharp Cheddar, Chevre, Gorgonzola, Camembert

Other names for Mourvèdre: Mataro in Australia; Estrangle-Chien, which translated is "dog strangler," in parts of France. Why dog-strangler? Hell, I don't know…never try and understand the French. And one up for debate: Monastrell. Monsastrell has forever been linked as just another name for Mourvèdre, but recent research suggests that it is genetically a different grape altogether. For our lesson though, they are close enough in varietal character to still lump them together for now.

Some final thoughts on these wines:
Try to focus on the fruit components as you do this exercise, knowing full well that everything else changes according to region, climate, and winemaker influence. And for this exercise in particular, regional variation is key. The difference between New World and Old World styles of Syrah is pronounced. Stylistic difference applies to the Mourvèdre/Mataro as well, with a lot more experimentation and varietal examples being provided by the New World.

These two grape varieties are more often than not components of blended red wines in Spain and France, although varietal expressions are increasing there as well. Classic varietal expressions of pure Syrah are already available are the wines of Hermitage, Côte Rôtie, and Cornas in the northern Rhône region…but you will have to dish out a pretty penny to taste those! While you can pay a ton of money for extremely well-crafted Australian Shiraz, you can also find a ton more solid examples for not a lot of money. Have at it.

Additional barrel flavors you may detect in these wines depending upon how much time it spent in oak: (From Lesson 4: Got Wood? The Influence of Oak)
> *Specific to these red wines:* Almond, bacon, burnt, cashew, cedar, coconut, nutty, pencil shavings, sawdust, smoky, spicy, toast, toffee, vanilla

➤ Woody descriptors: Cedar, charred wood, cigar box, coffee, green wood, oaky, pencil shavings, resinous, sandalwood, sawdust, smoky, toasty, tree bark, vanilla.
➤ Barrel flavor (malolactic fermentation) descriptors: Almond, banana, burnt caramel, butter, buttered popcorn, butterscotch, cashew, coconut, lanolin

Got it? Then get busy drinking those Rhône Rangers! Hi Ho Silver! Bottoms up!

	Comments
Syrah (France)	
Color:	
Bouquet (smell):	
Taste:	
Body:	
Compare/Contrast to other wines:	
Shiraz (Australia)	
Color:	
Bouquet (smell):	
Taste:	
Body:	
Compare/Contrast to other wines:	
Mourvèdre	
Color	
Bouquet (smell):	
Taste:	
Body:	
Compare/Contrast to other wines:	

Lesson 16: Americas' Finest: Zinfandel & Petite Syrah & Malbec

What they smell like. What they taste like. Where you get them from. What foods bring out the best in them. And many of the aliases that they go by.

Look! Up in the sky…it's a bird, it's a plane….no, it's our American Super-Grapes! And don't be so egocentric about the vicinity of your habitation—I'm talking about continents here, not countries; North America and South America to be precise. This lesson will deal with a tri-fecta of truly titillating varietal wines which span the Western Hemisphere! The Americas' finest reds on display.

But wait: don't be confused by my construal: the Americas produce fantastic examples of Cabernet Sauvignon, Merlot, Syrah, Pinot Noir and even Mourvèdre. So how can I pick these three varietals as 'Americas' Finest' reds? Mostly because of this simple fact: they aren't really produced in great number anywhere else! Zinfandel, Petite Syrah, and Malbec are three grapes that produce three wines which find their greatest expressions (both in terms of variety and of quality) in North and South America. They are grapes with no huge production areas or long historical pedigrees in Europe, but man oh man are their American experiences exceptional!

So break out your cape and prepare to go up, up and away with these super American reds…come on Clark Kent, let's drink this now!

What to grab:
- One bottle from each list of the following from your local wine shop or grocer:
  Zinfandel
  - **Bottles $5-$15**
    - ➢ Bogle Old Vine Zinfandel (California)
    - ➢ Flying Winemaker (California)
    - ➢ Cline (California
  - **Bottles $16-$30**
    - ➢ Sebastiani (California)
    - ➢ Cline Ancient Vine (California)
    - ➢ Ravenswood (California)
    - ➢ Bonny Doon Cardinal (California)
    - ➢ Ridge Geyserville (California)
  - **Bottles $30+**
    - ➢ Grgich Hills (California)
    - ➢ Ridge York Creek (California)
    - ➢ Rafanelli (California)
  Petite Syrah
  - **Bottles $5-$15**
    - ➢ Concannon Limited Release (California)
    - ➢ Bogle (California)
    - ➢ De Bortoli DB (Australia)

> Rosenblum Heritage Clones (California)
Bottles $16-$30
> Connor Park Durif (Australia)
> Peachy Canyon (California)
Bottles $30+
> Lolonis "Orpheus" Redwood Valley (California)
> Caliban (California)

Malbec
Bottles $5-$15
> Alamos (Argentina)
> Maipe (Argentina)
> Cedre Heritrage Cahors (France)
> Dolium Malbec (Argentina)
Bottles $16-$30
> Luigi Bosca (Argentina)
> La Posta (Argentina)
> Lagarde Single Vineyard (Argentina)
> Bodega Catena Zapata (Argentina)
Bottles $30+
> Monteviejo Lindaflor (Argentina)
> Bodegas Catena Alta (Argentina)

Don't worry if you can't find these exact labels. You can always ask your local wine store salesperson to help you pick out a good representation of the varieties I have described above. Show them the list and tell them to get you something close to it.

What to do:

1. First open the bottle(s) and try immediately. What is the color, the aroma, the body, and the taste of each wine? Record all impressions. Write down some descriptions of your experience, both what you smell and how it tastes. You should be able to detect some subtle difference in flavors and aromas in these wines…and perhaps even different colors as well. Really work hard at picking out and describing differences between each of these varietal wines. As always, include as many descriptors as possible to define each wine, even if some may be a stretch. Check below for a bit of varietal vocabulary to choose from.

2. Next wait at least 2 hours for the wine to "open up," leaving the cork off the wine to allow oxygen to get in. Try the wine again. Has the smell changed? How so? What about the taste? Is it fruitier? Does it taste better to you? As you go back and forth from one wine to the next a few times, be sure to cleanse your palate with a bit of bread or plain crackers. Sip some water as well, if you absolutely must. But don't eat any other food yet—it will easily alter the results of your descriptions. Don't worry: you can finish the bottles with dinner later.

3. Now let's revisit Lesson #1: try the wines with cheese or any appropriate food. Record your reactions. Has anything changed

in the taste? Does the wine seem smoother or lighter or fruitier? Repeat with a different type of food if available. Again, detect some differences and make yourself write down some different descriptors.

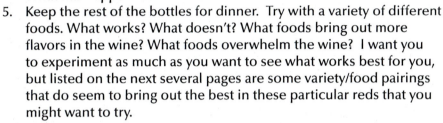

4. Finally (if possible), save a bit of wine in the bottle. Let it sit exposed (no cork) on the counter for a few more hours. Let it go overnight if you want. Then try it again on its own, without food. What's it like now? Write it down. Compare this with your first impression of the wine right after you opened it. What has happened?

5. Keep the rest of the bottles for dinner. Try with a variety of different foods. What works? What doesn't? What foods bring out more flavors in the wine? What foods overwhelm the wine? I want you to experiment as much as you want to see what works best for you, but listed on the next several pages are some variety/food pairings that do seem to bring out the best in these particular reds that you might want to try.

What to look for:

The whole point of this exercise, and others in this section, is to get you fluid and fluent in recognizing and describing the major varietal wines produced in the world today. Of course you know that the term *varietal wine* means: a wine made predominately, if not exclusively, from one type of grape. As you do the tasting, focus on comparing and contrasting these different wines, with more of an emphasis on contrasting.

The last tasting exercise in this book will compare some varietal wines from different regions, but you should do them on your own as often as possible too. When purchasing a bottle of one of these wines for dinner, go ahead and grab an inexpensive version of the same grape from a different part of the world. Your knowledge will grow exponentially, even if it is at the expense of your wallet decreasing mathematically.

This is actually quite a challenge for these varieties, as they are not widely produced around the world. Malbec varietals are mostly from South America, Petite Syrahs mostly from North America, and Zinfandel is almost exclusively a California affair. But you can certainly find lots of examples from different wineries, at lots of different quality levels, even within a single country or region. Isn't that super?

The Specifics:

Let's take a look at each of these major American varieties one at a time to provide some tasting and smelling clues, some regional background, and also some food paring hints. Let's start with…

Zinfandel

Like Superman himself, Zinfandel may be from another planet, but he came to recognize his true power during his American experience! And he is the undisputed heavyweight of the American superhero/wine pantheon. There are lots of other competitors, but there can be only one number one! Stand aside Batman and Green Lantern, and Shiraz and Cabernet Sauvignon...even you cannot match the strength and stamina of America's finest: Superman and the super-grape Zinfandel!

What in the hell am I suggesting with this super-analogy? Just this: Zinfandel is perhaps the only great red wine produced on the planet without an European pedigree. It's not grown anywhere in the Old World because they don't even know what it is! Some say it's a grape called Primitivo from Italy, others say it has Croatian roots. Either way, there is no great history of Zin production in Europe, no regions which specifically grow it, no definite region as a source of origin, no experimental Zin wines...anywhere. They have no experience with it, and don't seem keen to attempt growing it anytime soon either.

And quite frankly it wouldn't matter if they did. Apparently Zinfandel is one of those grapes that only found its best growing conditions after it traveled across the Atlantic and came ashore in America. We are pretty sure that the plant material we now call Zin was experimented with and planted on the East Coast of the US over two hundred years ago. But even then it didn't amount to much. It took the additional western trek across our great country before Zin found its true home in California. And just like Superman from Krypton, Zin's powers became exponential once it found a new home...

From the era of the Gold Rush onwards, Zinfandel has played a major role in the Californian, and thus American, wine experience. It grew prolifically in the hot California sun, and it was first produced as a bulk red wine for the masses—becoming the American version of a basic Bordeaux claret. In the 1980's, massive overproduction and surplus of the grape led to experimentation and evolution of 'blush wines' which boosted its popularity to American consumers' insatiable demand for light, fruity, semi-sweet drinks. In other words, White Zinfandel and blush wines were born...which of course paved the way for the bastard wine blasphemy that we know today as 'wine coolers.' Hasn't anyone been arrested or charged with that heinous act yet?

But I digress as usual. Zinfandel has had a very unique history of continual evolution into all products imaginable in the wine world. White, pink, or red...it's an extremely diverse grape which is made into a variety of styles. But that is not the real point here. If it simply was made into adult Kool-Aid, I wouldn't be talking about it in this book.

Here is the real deal: it is a true king in the wine world. Bold, assertive, tannic, full of color, full of flavor, full of life...Zinfandel can be crafted for longevity, for complexity, and for greatness. And it often is. It is the American equivalent of Bordeaux's Cab, of Burgundy's Pinot, and of Barolo's Nebbiolo. But wait, those are Old World kings...and we Americans threw off the yoke of monarchy...so perhaps we should refer to this grape as President Zin!

And the President does not disappoint! Over the top jammy, raspberry flavors combine with a herbaceous smokiness which not only makes the wine distinct, but ultimately complex and intriguing. And the grape grows so damn well in the sunny Golden State that the berries accumulate huge amounts of sugar, which as you well know by this point in the book equates to high amounts of alcohol! This stuff is the bomb...almost literally! Zin often achieves alcohol levels in the 15 to 16% range, which is about the upper absolute limit for wines!

Big body, big tannins, big color, big aromas, big flavors, and big alcohol! Wow! If that ain't American, I don't know what is! You can enjoy these guys young, but well-crafted ones will become even more expressive of their super-powers with age.

So grab a great Zin and ...Up, up, and away!

Distinct Zinfandel descriptors: Jammy, red raspberries, briary-rhubarb, berry-smokiness

General Zinfandel descriptors: Black pepper, bramble berry, briar, cherries, cinnamon fruitcake, herbal, plum pudding, prunes, raspberry, red pepper, rhubarb, ripe berries, spicy, wild berries, earthy, gamey, nutmeg, savory, tar, truffle, cashew, cedar, coconut, nutty, pencil shavings, soy, vanilla, wood smoke.

Where Zinfandel is grown: California, California, and California. Oh, and California.

What foods are often paired with Zinfandel: Mild to strong to very strong cheeses, garlic rubbed tangy spareribs, lamb casserole, pizza, calzone, steak, rib roast, barbecued ribs or chicken, wild boar or venison prepared anyway possible, spicy sausages, eggplant lasagna in a spicy tomato sauce...wow I'm getting hungry, grilled steak, grilled pork chops, barbecued pork chops, roasted game birds, chili, cheeseburger with jalapenos peppers, cheese-steak hoagie, steak calzone, blue-cheese covered hamburgers. Okay I've got to go eat now... Hey man, this is the distinctively American red, so match it with hearty, distinctively American foods!

And let's do some specific cheeses too
Definitely do: Monte Enebro, Bandaged Cheddar, Flagship Reserve, Dry Jack, Asiago
Will work: 1yr Cheddar, 2yr Cheddar, Grana Padana, Cranestrato,

Montbriac
Possibly try: Buffalo Mozzarella, Smoked Gouda, Smoked Cheddar, Chevrelait

Other names for Zinfandel: Primitivo… maybe.

Petite Sirah

Well, we are coming off the true American red bombshell, and so by shifting to something called 'petite' you are probably thinking that we are going to lighten the mood a bit…we can relax, and get ready to enjoy a lighter wine….a little-er wine…something nice, and delicate…something smaller…something petite….

Not gonna' happen! Wake up! Get the hell up! It's time to assault your palate again with another powerhouse red wine that has come to the New World for its fullest expression of audacity and strength! Petite Sirah is most often expressed as a dark inky black wine with significant acidity and tannins, which can give it enough chewy texture and mouth feel to choke a Chihuahua. This stuff is dark as night…it may be the darkest wine you ever encounter, and there is a reason for this that is related to its name.

See, there ain't nothin' petite about Petite Sirah…except for the size of the grape itself. The actual berry size is indeed small, which means there is a high skin to juice ratio when undergoing the skin contact phase of red wine production. (Don't remember what that means? Check back to Lesson 6: Color Me Clueless.) This situation can produce very tannic wines of *extreme* dark colors if the juice goes through an extended maceration period. In addition, when Petite Sirah gets aged in new oak barrels for any length of time, the wine can develop an aroma of melted chocolate. Hints of chocolate already in the red wine? Damn! This wine is just super!

The end result: big American wine. With herbal and black pepper overtones on top of a dark berry flavor, Petite Sirah may be related genetically to the Syrah grape, but it is distinctly darker purple/ blacker and much bolder than its relation. Bold is the operative word here. Bold color, bold tannins, bold acids. Add those three factors up and you get a heavy bodied wine with chewy mouth feel, and extreme aging ability. These wines can easily take ten to twenty years to mellow out enough to achieve their full potential.

If Zinfandel is Superman, then consider Petite Sirah as your Batman…the Dark Knight of wine!

Distinct Petite Sirah descriptors: Black currents, blackberry, boysenberry, black berries of all sorts, black tar, black pepper, black as night, is it black enough yet? And maybe even a hint of dark chocolate as well.

General Petite Sirah descriptors: Bramble berry, briar, herbal, chocolate, dark berries, earthy, fruity, inky, licorice, mulberry, plum, raspberry, dusty, earthy, leather, mushroom, prune, spicy, oaky, tobacco leaf, vanillin, smoked meat, candied plum, bitter chocolate, espresso, blackberry liqueur

Where Petite Sirah is grown: Again, mostly California. However, Australia does produce a bit under the label of Durif. And look for interesting examples coming out of a truly bizarre grouping of minor wine-producing countries: Argentina, Brazil, Mexico, and Israel.

What foods are often paired with Petite Sirah: Mesquite grilled steak, grilled lamb, grilled duck, roasted duck, pot roast, grilled rabbit, sautéed rabbit, rabbit in mustard sauce, beef stew, barbecued pork, barbecues ribs, barbecued anything, roast chicken in any heavy sauce most notably garlic and lemon, hearty pasta dishes. Deep red sauce dishes. Any beef dish, steak prepared just about any way you want…but the more meaty flavor to come through, the better.

And let's do some specific cheeses too
Definitely do: Chimay Grand Cru, Goat Gouda, Soumaintrain
Will work: Mahon, Arina, Benning, Darcy, Pierre-Qui-Vir
Possibly try: Gorgonzola, Manchego, Cheddar (sharp)

Other names for Petite Sirah: Durif in Australia. You might also see reference to Peloursin, which along with Syrah is the parent genetic material for Petite Sirah. Peloursin is a distinct varietal, but is almost indistinguishable from Petite on the vine or in the wine.

Malbec

Now we head south of the border to investigate our last big red in the American line-up… all the way to South America to be exact. As Zinfandel is to California, Malbec is to Argentina: it is the premier grape for the region that is not produced widely as a varietal wine anywhere else in the world. However, if my powers of prediction are on mark, it soon will be… 'cause it is awesome!

Previously Malbec was known only as a very minor varietal of Bordeaux which was used exclusively in small amounts as a blending grape; it is one of the 'big 5' grapes used in the Bordeaux blend, albeit in trace amounts… if at all. However, much like Zin's travels to sunny CA, Malbec has found a new home in South America. This grape grows prolifically on the sunny slopes of the Andes Mountains in Argentina and Chile, and produces a rich, supple wine which would almost be unrecognizable to folks back in its European homeland.

I believe we will see more and more examples of Malbec popping up around the world, in direct correlation with the subsiding popularity of Merlot. In other words, Malbec may be the new Merlot! And that is an appropriate comparison: Malbec is already commonly referred to as, 'a softer, rounder little brother to Merlot.' I agree and disagree with this description. Malbec does have many of the plum, cherry flavors and aromas that Merlots do, and it is subtle, round and soft. But Malbec has so much more going on than the average Merlot does…

Well-crafted Malbec wines are soft and silky even when young, but the flavors are much more pronounced and integrated. They also have a degree of depth and character which is lacking in the ocean of mediocre Merlot. Malbec often has a rich, heavy mouth feel with a soft silkiness normally associated with lighter wines…which makes them kind of unique. And here's a heads up: With even a little oak barrel aging, even average Malbecs gain a backbone and complexity that makes them intriguing and delicious!

Here are a couple more Malbec bonuses:
1) Unlike its American brothers Zin and Petite, Malbec is accessible even in youth. Its fruity flavors and tannins are seemingly softened immediately upon the wine's creation. With the addition of barrel aging, Malbecs become even softer and silkier…and more complex. What a delight!
2) Because they come from South America, the prices are significantly cheaper! Sorry to all my Latin American cohorts, but you guys know it's true! Things are cheaper down there, and the result is that quality South American wines are much cheaper than equivalent quality wines from Europe, or even California. You can get some total kick-ass barrel-aged reserve Malbecs that are simply outstanding for 20 to 40 bucks. That sound like a lot to you? Go around the globe and see how much more quality wines that have significant barrel aging cost…you will soon figure out how sweet those South American wine deals truly are!

Great Malbec wines, at great prices…man, I love Argentina and Chile more and more everyday! For you novice wine drinkers, if you want to sample high quality wines at a fraction of the European costs, look no further than our South American neighbors. And specifically their American super-grape, Malbec!

Distinct Malbec descriptors: Ripe plum, wild violets, hints of cherry, with a soft silkiness even in young wines.

General Malbec descriptors: Plummy, violets, floral, cherry, black cherry, black currents, blackberry, boysenberry, black and red berries of all sorts, chocolate, earthy, mulberry, plum, raspberry, earthy, light leather, prune, oaky, tobacco leaf, vanillin

Where Malbec is grown: Argentina, Argentina, and Argentina. Malbec is Argentina's premier wine grape. Major grape of Chile as well, from which I have personally had some spectacular reserve Malbecs. In France: Bordeaux of course, where it is used very sparingly as part of the Bordeaux blend, some in the Loire Valley, and the 100% Malbec wines from the small French region called Cahors is always an inky black intense wine experience (it is southwest of Bordeaux). A bit is grown in California and Australia, but again mostly for blending purposes. Stick with South America for your best bet.

What foods are often paired with Malbec: Pancetta is the bomb, salami, sweet hot or spicy sausage, grilled pork chops, grilled veal, sautéed veal, roasted veal; Mexican dishes like beef tacos or burritos or chili rellenos; Cajun dishes like gumbo and jambalaya; pizza and calzones of all types; pasta and ravioli in meat, vegetable and tomato sauce; chicken cacciatore or chicken in mushroom sauce. Or just a big ol' rare grilled steak. This wine loves meat, from picnic cold cuts to sizzlin' hot on the spit.

And let's do some specific cheeses too
Definitely do: Morbier, Asiago, Iberco, Taleggio
Will work: Cashel Blue, Maytag Blue, Manchego
Possibly try: Cheddar (sharp), Camembert

Other names for Malbec: Known as Côt or L'Etranger or Pressac in different parts of France. And sometimes labeled as Malbeck in Argentina.

Some final thoughts on these wines:
Try to focus on the fruit components as you do this exercise, knowing full well that everything else changes according to region, climate, and winemaker influence. But with these wines, regional variation is not as key since so many of them are exclusively American affairs. While all three of these varieties are big and bold, you still should not have any trouble telling them apart in terms of different tastes…they are pretty distinct powerhouses each in their own right.

To recap a bit:
> Zinfandel and Petite Syrah varietal wines are almost exclusively produced in the US. They both have incredible color extraction and body and mouth feel.
> Zinfandel is extremely diverse in that it is made into a variety of colors and styles of wine. However, its role as a big red titan of America is what we focused on for this lesson. For all its big characteristics and aging potential, Zin can be considered one of the Kings of the red wine world…and the only one lacking a European pedigree.
> Malbec is sparsely used as a blending grape in parts of France, but is almost exclusively made into a varietal wine in Argentina and Chile. With the exception of a sub-region of Bordeaux named Cahors which produces a 100% Malbec.
> South American wines, and I'm speaking primarily of Argentina and Chile, are significantly cheaper than their European or American counterparts. This is especially true for higher quality wines that have significant barrel aging.

Got it? Then continue your exploration of the Americas by drinking new discoveries of Zinfandel, Petite Syrah, and Malbec. Columbus would be proud!

	Comments
Zinfandel	
Color:	
Bouquet (smell):	
Taste:	
Body:	
Compare/Contrast	
to other wines:	
Petite Sirah	
Color:	
Bouquet (smell):	
Taste:	
Body:	
Compare/Contast	
to other wines:	
Malbec	
Color	
Bouquet (smell):	
Taste:	
Body:	
Compare/Contrast	
to other wines:	

5.
STYLISTIC DIFFERENCES, SIGNIFICANT OTHERS, & YOUR JUST DESSERTS...

And now for the last. Hopefully by this point in the book, you have sampled dozens and dozens of wines, learned a bit of the winemaker craft, and understand the components of wine enough to converse with any fellow imbiber about the quality and style of any wine you are sipping on at a cocktail party. Outstanding initiative! I'm glad you made it! Welcome to the wine world!

You are now well-equipped enough to handle just about any wine experience that comes your way. But I didn't want to end this party just yet. I have at various points in these lessons referred to 'Old World' and 'New World' styles of wine, suggesting that there is some distinct difference between the two. Let's take one more lesson to set up a series of comparative tastings which highlight these major stylistic differences in our wine world. Then we'll wrap this baby up with a couple of chapters that aren't so much lessons as lists of off-the-beaten path wine alternatives...lesser-known grape varieties and wines which book-end a meal as the apéritifs and desserts.

First, the New versus the Old: let the battle begin... And then get your just desserts in dessert wines and alternative grapes. Party on!

Lesson 17: When Worlds Collide: New vs. Old in a Stylistic Showdown
Lesson 18: Significant Others: Up and Coming Varieties
Lesson 19: Before & After the Meal: Apéritif and Digestif Wines
Epilogue: The End is Only the Beginning

Lesson 17: When Worlds Collide: New vs. Old in a Stylistic Showdown

Difference between 'Old World' style and 'New World' style wines. What that means. How it got to be that way. Multiple comparative tastings you can conduct at your own pace.

Throughout this wine tasting lab manual, I have frequently referred to different wines from different places exhibiting either a 'New World' or 'Old World' style. If you have actually gone through this entire book, you probably now have a good sense of what these terms mean…but let's chat just a bit more about them in greater detail now. Because I really want you to understand what these terms are describing…and it's not simply a matter of different locations my friends. Oh no! It refers to differences in style, differences in focus, differences in techniques, differences in motivations, and differences in overall philosophies that the wine makers have. Wow! That is deep!

Deep indeed my friends. The Old versus the New is a struggle that spans the millennia of human experience. Tradition versus innovation. Conservative values versus openness to change. Aged wisdom versus youthful enterprise. And so it is with wine…

Old World, Old School

See, a thousand years ago and five hundred years ago and even up to fifty years ago, almost all wine consumed across planet earth came from Europe. Europe has forever been the laboratory for the evolution of wine grapes, of wine styles, of wine technology, and of wine industries. It is no coincidence that all the most recognizable and most planted grape varieties in the world all trace their origins back to Europe. And Europe, or the 'Old World' countries of France, Italy, Spain, Portugal, Austria, Germany, still produce the lion's share of exported wine even today.

In the Old World's wine evolution, geography became of the utmost importance not just in defining the wine, but also in the philosophical outlook of wine production. Let me explain. In the Old World, winemakers consider the wine they produce an expression of the place it is made. You will often hear this term referred to as terroir, a mystical French term which has no exact translation to any other language. But I will give it a try…

The *terroir* of a place is a combination of all the physical and cultural factors which are unique to every space on the planet. Meaning, each plot of land has a unique soil, unique rainfall, unique sunlight, unique humidity, and every other thing you can think of. But that's not all. *Terroir* also has cultural dimensions, which means that the human decisions for the vineyard also play a role in its uniqueness: the way they prune the vines, the fertilizer they use, the type of grapes they grow, the way they make the wines from those grapes, etc. All of those physical factors plus all of those human factors combine to make a unique *terroir* for each vineyard/town/region. To them, this equates to a unique type of wine, one that is distinctly driven by place.

And that is why they name their wines according to where they are from, not by the grape of which they are made. Make sense yet? You may have heard reference to a Burgundy, a Chianti, a Bordeaux, a Barolo, or a Rioja wine. Those are all places man! The *place* expresses what the wine is. If you lived in Europe, you wouldn't have to be told that a wine from Burgundy is 100% Pinot Noir...everybody just knows! That's the way its always been! Not only is each region unique, but even each town is unique...hell each vineyard is unique! That's why they put so much damn information on the wine labels in Europe: they are telling you more and more about exactly where the wine is from to further and further point out how unique it is!

But it's even deeper than just a wine labeling convention; it is a true philosophy of how to create the wine itself. Old World European winemakers typically are crafting a product that best expresses the place. They are basically stewards that are simply there to help the land and the grapes express their individual character to create this unique wine, from this unique place. That's why you may sometimes see them referred to as *wine growers* as opposed to *wine makers*...does that make sense?

Expressing this uniqueness—and all the recipes that have evolved for blending the wines—is a matter of tradition that has been passed down over the centuries in most European wine regions. History, tradition, geography...wine. And wine that is crafted to go along with the local cuisine: which is expression of the local culture, unique to every place and region too. To Europeans, wine is food; matching the local cuisine to the local wine is also a deeply entrenched part of their philosophy. Go back and read Lesson #1 again: its why Old World wines best express themselves with food. Starting to get the picture here?

All this wine history and wine tradition has manifested itself into strong regional identification with a specific wine style. And this in turn has resulted in strict laws which dictate exactly what is grown, where it is grown, how it can be blended, and how it is labeled. You may grow fantastic Pinot Noir in France, but if you are not located in the officially designated region named Burgundy, then you can't label it Burgundy. And even if you are located in Burgundy but you like to blend in some Merlot into your Pinot Noir, then you ain't following the rules, and therefore cannot label your wine as a Burgundy.

So in the Old World the rules are set, the varieties are all matched up to the places they grow best, the recipes for blending are tried and true. The wines are crafted to match the foods, which means that many focus on earthiness and spiciness as opposed to just expressing the grape flavors. And many have a tannic or acidic structure which cuts into fatty foods in order to uncover the subtle flavors of the wine underneath. This tannic component also makes their wines more age-worthy…and sometimes forces aging before being accessible at all.   History…tradition… geographic expression…old school, old world wines…

But then Columbus bumped into a big-ass landmass as he sailed west…

New World, New School
Enter: the New World. All those ex-colonies of the European countries that threw off the yoke of monarchy and started something new. And the same goes for their wine producing experiences too…

See, the New World was, in a word, new. It didn't have a rich wine growing history. It didn't have the wine growing grape evolution. It didn't have a tradition of matching wine with local foods. It didn't have a distinctive local cuisine. It didn't have strict laws which dictated what grapes were to be grown and what recipes were to be followed. In fact, there were no wine grapes or wine rules at all! Now that I think about it, they didn't have much wine consumption either, since all of it had to be imported from Europe; a costly proposition back in the day.
I don't mean to go into great historical depth here, so I will fast-forward the story. As the New World wine industries of The United States, Argentina, Chile, Australia, New Zealand and South Africa evolved in the last two hundred years, they initially took their cues from the Old World masters. They imported the wine grapes from Europe, and even spent most of their time trying to mimic European wine styles and wine blends.

But, you can rarely duplicate a wine when you are growing the grapes in a radically different climate and radically different soil…to sum up, in a radically different *terroir*. And some grapes wouldn't grow at all in the new places (think Nebbiolo), while others proliferated better than they ever did back home (think Zin). Which eventually led the New World wine growers and wine makers to experimentation. And that my friends is the start of the separation of styles.

Experimenting with new varieties, with new blends, with new styles altogether became a hallmark of the New World wine style. And since the New World-ers were not constrained by laws or by tradition, anything goes! The Americans in particular took the initiative and radically increased the use of lots of new wine technologies that they themselves invented. Things like: the use of refrigeration during fermentation, the use of stainless steel tanks, the selection of specific wine yeasts, and genetic selection and manipulation of the grape plants themselves. New technologies and new techniques, which led to new wine styles…

New wine styles which were reflective of the New World philosophy on wine production: a philosophy focused on wine as a product of the wine maker that makes it, not as an expression of the place it is from. And in the world of wine, that was as revolutionary as a democratic republic! In the New World style, the wine maker isn't trying to express 'Napa' or 'New Zealand'; he is more often than not doing his best to express the Chardonnay grape or his own particular brand of Chardonnay that is heavily manipulated to create assertive flavors that he thinks people will like.

I suppose in a way it is a very democratic form of wine making...to each his own expression. And New World wines were flexible enough to cater to market demand as well, so when a wine maker stumbled across a flavor or style people liked, then he went with it! This is a process that is still going on today. Lots of experimentation, lots of market input, and lots of catering to market demand. So what has the New World market demanded? Let's stick with the American experience for now...

As wine novices (as many of you were before you went through this book), the American palate naturally gravitates towards simple, light, fruity wines with not too much complexity. Remember, the American beverage palate is all about Kool-Aid, sports drinks, soda pop and light beer; we like it simple and sweet. We also have a long history of consuming alcoholic beverages with the sole purpose of inebriation...there is no entrenched culture of consuming alcohol with food, as a component of the meal. Therefore, we also like wine to be accessible to our tender palates right now; we like it to be up front; we like to be able to sip it without having to eat any food in order to transform the wine.

Add the lack of wine history with the tendency towards experimentation with the desires of the American palate and...presto!...varietal wines are born. Wines made solely from one type of grape, and heavily manipulated by the winemaker for style. They can also be known for being heavily oaked, or heavy with vanillin, or over the top with fruit flavor...all traits expressly crafted by the winemaker according to their personal preference or market demand. New World style wines are typically referred to as 'fruit-forward' meaning that the first thing that you recognize is the fruity components of that particular grape.

And most of the time that is exactly the point. Chardonnay, Cabernet Sauvignon, Merlot. That is what it says on the bottle; that is what is inside. The wine maker is expressing the varietal characteristics of that grape... not a focused expression of the place it is from. Think about it. In the New World, 'Napa' or 'Oregon' is not indicative of a particular grape, a particular blend, or a particular style. It's just a place a lot of different wines are made. Wines which mostly express the essence of grape or the winemaker's skills....or both Get it? Got it? Good.

The Result

Put all this together now to understand the really important manifestations in the wine of these two radically different styles:

❖ *Old World-style* crafted wines are typically much more acidic and tannic, because they are built to be matched with food. Old World wines really only express all of their intricacies of character with food. As you hopefully learned in Lesson #1, Old World wines typically have earthy and spicy undertones alongside the fruity aromas and flavors added by the grapes. The higher tannic/acidic components also means they are built to last, and integrate flavors over time. Old World is also more prone to using different grape varieties which they blended together to achieve a certain style/taste (thus, most Old World wines are blended wines.). Old World emphasizes wine as the expression of the place: therefore wine regions, wine sub-regions, wine villages, and even individually named vineyards all represent distinct wine styles or distinct levels of quality...or both.

❖ *New World-style* are typically more fruity, more creamy, more vanilla, or more soft so they are more approachable and drinkable upon creation. Even ones with deep color and high tannic structure are intentionally softened with a lot of oak contact, so that they are accessible, even in youth. New World is crafted for now. More importantly, New World indicates an emphasis on the individual winemaker or the individual expression of the grape itself. Extreme experimentation can drive some New World styles to the edge of over the top jammy flavors, over the top alcohol or over the top body...but all somehow still immediately drinkable of course. New World wines are not trying to exemplify a unique expression of a village, a region, or a state. New World wines are therefore typically focused on expressing the best of the fruit, or in expressing the desires of the winemakers....or both.

So let's taste this difference now

The following pages contain tastings of a single varietal or blended wine, with lists of examples from the opposing styles that you can use to set up your own little cockfight with. Not all at once of course, but do a comparative tasting with one of these varietals/blends as time permits in your life. Please keep in mind, that there is no definitive winner here. There is no style that is 'better' than the other: it's just different. Which of course makes the world of wine, and all of us in it, much richer for the diversity of choices.

As with all the tastings in this manual, make it a social event...invite over friends or family, and have them all bring an example of that single type of wine, but from different places, and thus of different styles. You are schooled enough to know how to conduct a comparative tasting at this point with no further instructions from me. I would only suggest that you take good notes, recording your impressions of the differences you detect between the two styles. And I simply must insist that you do this as a social activity. Not only is it way fun-er, but the conversation that is generated is the greatest part of the learning experience! Wine is a social drink, and its greatest power is not individual pleasure but the communion that is created by sharing and enjoying it with others.

But enough of the sappy shit! This lesson is about struggle, about competition, about who will come out on top after this showdown concludes! Are we ready? Then let the battle begin! Let's drink this now!

Cage Match #1 Chardonnay
New World Chardonnay
  - ➢ Acacia (California)
  - ➢ Benziger Reserve (California)
  - ➢ Alexander Valley Vineyards (California)
  - ➢ Sonoma-Cutrer (California)
Old World Chardonnay
  - ➢ Louis Jadot Macon Villages (France)
  - ➢ Domaine de la Cadette Bourgogne Vezelay (France)
  - ➢ Ramonet Vergers Chassagne Montrachet (France)
  - ➢ Domaine Pinson Les Clos Grand Cru (France)

What to look for: New World Chards are known for heavy barrel flavors or being 'rich and oaky'. Heavier on the palate, made from very ripe to over-ripe grapes which add more alcohol, more body, and more fruit flavor too. The heavy-handed use of oak (meaning the wine spent more time in oak, in newer oak, and typically American oak…all of which pack more of a punch) results in more oaky, toasty, coconut, vanilla or butterscotch. Old World Chards are typically light, crisp, refreshing and subtle on the fruit flavors. Mostly un-oaked or lightly oaked, these light to medium body wines sometimes have a steely acidity and hints of minerality.

Scuffle #2 Sauvignon Blanc
New World Sauvignon Blanc
  - ➢ Ferrari Carano Fume Blanc (California)
  - ➢ Whitehaven (New Zealand)
  - ➢ Kim Crawford (New Zealand)
  - ➢ Cakebread (California)
Old World Sauvignon Blanc
  - ➢ Caves des Perrières Pouilly Fumé (France)
  - ➢ Domaine Joël DelaunamTouraine (France)
  - ➢ Domaine Reverdy Sancerre Cuvée "Les Coutes" (France)
  - ➢ Vincent Pinard Sancerre (France)

What to look for: New World Blancs, like their Chardonnay brothers, are known for barrel flavors or being 'rich and oaky', but more so for over-the-top distinct fruit attributes. Cat pee accentuated? Are you kidding me? No I am not. Medium to heavy in body, made from very ripe grapes which add a bit more alcohol, more body, and more of that distinct Blanc grassy fruit flavor too. The use of oak (meaning the wine spent more time in oak, in newer oak, and typically American oak…all of which pack more of a punch) results in more oaky, toasty, coconut, vanilla or butterscotch, but usually not as over the top as many Blancs.

Old World Blancs are typically dry, light, crisp, refreshing and subtle on the fruit flavors. Mostly un-oaked or lightly oaked, these light to medium body wines sometimes have a crisp acidity and hints of minerality. The Blancs can sometimes achieve a steely acidity that has real 'bite' potential.

Street Fight #3 Pinot Noir
New World Pinot Noir
> ➤ Tohu Cuvee (New Zealand)
> ➤ Castle Rock Mendecino (California)
> ➤ Stephen Vincent (California)
> ➤ St. Innocent Shea Vineyard (Oregon)

Old World Pinot Noir
> ➤ Domaine Nicolas Rossignol Bourgogne (France)
> ➤ Domaine Alain Hudelot-Noellat Bourgogne Rouge (France)
> ➤ Domaine Lecheneaut Nuit St Georges Les Damodes (France)

What to look for: New World Pinot Noirs are known for being the most fruit forward Pinots on the planet. While barrel flavors are incorporated, these wines are well-integrated even when young, and just really express the fruitiness of Pinot like no other. The more money you spend, the richer, cherrier, plumier flavors you will get right off the line. And smooth, silky and fine even when young.

Old World Pinots mostly from Burgundy are a different animal altogether. When young, the tannin-acid bite almost demands a food paring. More importantly, they lack the high level of fruit flavor that their New World counterparts exhibit. In the place of fruit is a savoriness, a dryness, and undertones of earth, barnyard, and spice.

Mosh Pit #4 Malbec
New World Malbec
> ➤ Maipe (Argentina)
> ➤ Dolium Reserve (France)
> ➤ Colome Estate (Argentina)

Old World Malbec
> ➤ Chateau du Cedre Cedre Heritage Cahors (France)
> ➤ Georges Vigoroux Gouleyant Malbec (France)
> ➤ Chateau Lagrezette Malbec Cahors Le Pigeonnier (France)

What to look for: New World Malbecs are all about the fruit-forwardness, and it won't be hard for you to pick it out. On top of that, integration of oak flavors from the barrel almost immediately marry into the body and flavors of the wine and take it to the next level of complexity and smoothness...while somehow still maintaining that great red berry fruit flavor.

The Cahors Old World style Malbec is a black as night tannic bomb. When young, the mouth feel is chewy as cheese, and the body is heavy. The tannic/acid bite demands aging in order to give it time to mellow out, but young or old don't expect the same fruit flavors as its New World counterpart. So what do you taste? And how do those aromas and flavors differ from the Argentinean example?

Skirmish #5 Syrah/Shiraz
New World Shiraz
  - ➢ Milton Park (Australia)
  - ➢ Step Rd Blackwing (Australia)
  - ➢ Penfolds Bin 128 Coonawarra (Australia)
  - ➢ Thorn Clarke Shotfire (Australia)
  - ➢ Mollydooker The Boxer (Australia)
Old World Syrah
  - ➢ Chateau Grand Cassagne Hippolyte (France)
  - ➢ Chateau Guiot Costieres de Nimes (France)
  - ➢ La Reverence (France)
  - ➢ Copain Wine Cellars Eaglepoint Ranch (California)

What to look for: This one is the rumble in the jungle to beat them all! New World Australian Shiraz can be incredibly dark, heady, high alcohol, high body, low acid affairs that are heavily oaked to grab some heavy vanilla flavors. Loads of fruit flavor and firm but velvety tannins, right at the point of creation. A whole lot going on, but you can start drinking it right now anyway.

The Old World style Syrahs are actually not terribly different in most respects to the Shiraz descriptors listed above. However, tannic structure much more assertive in youth, and spice is definitely the trump card over the fruit. Not bombed with oaky vanilla hints, so not as creamy as velvety and easy as the Shiraz, but the complexity and depth comes out when matched with appropriate foods.

Battle #6 GSM Madness!
New World GSM
  - ➢ Rabbit Ridge Allure (California)
  - ➢ Kaesler (Australia)
  - ➢ Rosemount (Australia)
Old World GSM
  - ➢ Chateau Pesquie Terrasses (France)
  - ➢ Domaine Mathieu Chateauneuf-du-Pape (France)
  - ➢ Domaine Grand Veneur Chateauneuf-du-Pape '05 (France)

What to look for: The Grenache-Syrah-Mourvedre blend from the New World style mimcs the Old School for recipe, but not for complexity and depth. They can be incredibly dark, heady, high alcohol, high body, low acid affairs that are heavily oaked to grab some heavy vanilla flavors. Loads of fruit flavor combine with spicy hints.

The Old World style GSM is just the reverse: Loads of spicy flavor combine with fruity hints. More of those fruity hints will come on display when matched with appropriate foods. Tannic structure much more assertive in youth, and spice is definitely the trump card over the fruit.

Some final thoughts on these wines:

Psyche! This lesson was total hype! I don't believe in any of this Old versus New World bullshit! Okay, I'm kidding about that, but I wanted to get your attention because I need to throw in a serious disclaimer here:

Old World style versus New World Style **IS NOT** a geographic phenomenon anymore!

Huh? What the hell am I talking about now? Just this: both styles are now 'practiced' all across the world nowadays. Some winemakers in Europe have gotten tired of the constraints of their stringent system, and have started experimenting with New World styles. (Remember the 'Super-Tuscans'?) And some wine crafters in the US, Australia, and South America have started to gravitate back to experimenting with making the classic styles again. So some new blood in the Old School, and some old school skills popping up in the New World…but why?

It's easy to understand why the Europeans are now up for change: their creativity as wine artists has been stifled for decades, and the younger generation has had enough. They are bringing the vibrancy back into the business. And speaking of business, that is the other reason the Europeans are now mixing it up; they have been losing market share of their wine exports for decades to the New World, especially in the last ten years. France and Italy used to be the undisputed biggest exporters of wine forever and ever. Australia is now poised to dethrone them from this title. The Frenchies are more than a little miffed about this.

So miffed in fact that a major change is currently underway for the Europeans: they are not only crafting more and more wines in the accessible New World style, but have even softened their wine laws, and for the first time ever are allowing producers to label their wines by the grape variety instead of by the geography. That is major. You can see this transformation first-hand the next time you browse through the local wine store: look for Burgundies now labeled as 'Pinot Noir' or 'Merlot' from Italy. This is not to suggest that wine regions and regional blended wines are going to disappear—that will never happen—but a lot of catering to market demands is underway in the Old World.

In the New World, the move by some back to the traditional styles seems to have something to do with the maturity of their industries. Many wine makers are re-learning that classical wine heritage…and in staying true to form, are experimenting with it. Market competition is also at play: there are a million mediocre Merlots on the market right now, which has inspired some winemakers to craft something which stands apart from the pack: an Old World style Merlot. How hilarious this role-reversal has become!

On top of that, and as evidenced by the fact that you are reading this book, the New World wine consumers are maturing too! The US is becoming a wine consuming society at long last! As such, we are learning about wine, diversifying our palates, and many are yearning for more than the quick and easy, immediately pleasing, New World style. Complexity, depth and an evolving product are starting to become desirable. In addition, many many, many folks are starting to discover the true joys of wine and food pairings…a situation that is increasing the demand for those great Old World style blends.

So everybody is doing a little bit of everything, everywhere. But these two stylistic descriptions are still meaningful in understanding wines in today's world. Reading about or recognizing the style a wine is crafted in should tip you off to a host of other useful tidbits: what the wine is striving to be; what it is going for; can it age; should it age; should it be paired with food; or can you sip it on its own.

Drink it now or later? Have with food or solo? Sip or swill? All questions you can quickly answer after a single sip, which should tell you if the wine is Old World or New World style.

Lesson 18: Significant Others

Wherein we point out some red and white 'second-string' grape varieties that you might want to be familiar with, as well as some promising grape varieties that may have the temerity and tenacity to make it to the big leagues of the wine powerhouse players.

Now up to this point in this masterful manual of wine drinking, I have focused on the couple dozen grape varieties that are made into the most popular, most produced, most accessible, and most recognized wines on the planet. Indeed, you may not see any other wine options available in most wine shops or restaurant wine lists across the country. But hold the phone my friends! There are over 10,000 listed varieties just within the species *Vitis vinifera* (the "European" or "true" wine grapes)! This number is wildly disputed of course, with duplication of varieties and names occurring worldwide: one man's Primitivo is another man's Zinfandel...

However, wether the number is 1000 or 10,000, only 3 of these varieties are used for virtually all raisin production; about a dozen account for all table grape production....which leaves a really huge number of other varieties to be used exclusively for wine production. Of these, maybe 250 have some prominence in the world of fine wines and hundreds more are used for wine in limited or local areas; when you travel across Europe you will encounter hundreds of varieties produced as the 'local' wine in villages across the continent. Some of them have been used to make unique aperitifs and dessert wines, but most of these lesser-known varieties have been used as part of a winemaker's arsenal of tools to create blended wines. Rarely were they made into a stand-alone varietal wines like Chardonnay, Pinot Grigio or Merlot. And you just won't see a lot of them here in the US. But....

As the world of wine continues to globalize and more options are becoming available to wine drinkers the world over, we are starting to see lots of these hidden treasures start to surface...both from European regions that have been growing them for centuries, but also from New World regions that are much more open to experimentation and have no qualms about producing and marketing grape varieties that are otherwise unknown. New growing regions, re-emerging old wine regions, new marketing strategies, and increasingly sophisticated wine palates worldwide are driving this movement to try these other, lesser-known grapes and the wines they produce.

Thus: this! I wanted to provide you with a brief listing of what I consider some up-and-coming varietal wines that you will increasingly be bumping into at your local wine shops and restaurants in the future. Some of these grapes have already made a small name for themselves in the American market, and others will soon follow. They are all interesting second-stringers that I strongly suggest you get to know sooner as opposed to later when everyone else finds out about them...and thus the demand and prices will go up! None of them will ever be displacing Cab Sauv or Chardonnay; in fact most will never make it anywhere near the big leagues. But the fun of wine is continually expanding your palate and experience, and many expressions of these wines can really be interesting and sometimes fantastic. Give them a go!

No outlined or specific tasting exercise here. I mostly just want you to be aware of other wine alternatives that are out there. This list is certainly not exhaustive and is presented in no particular order, but tap into these significant other varietal wines when presented with the opportunity:

Significant Other Whites
Torrontés
The South American Pinot Grigio! With Latin flair! The most distinctive of all Argentine wines, because Argentina is pretty much the only country that produces it! Who the hell knows where it came from, and who really cares? It is considered a wholly Argentine variety now, even though Chile has started producing some too.

And it is an absolutely delightful white varietal wine! It has subtle hints of some very aromatically floral wines like Viognier and Muscat, with hints of mango, peach, orange, violets....but the aromas are always subtle, never over the top. The delicate aromas are almost always bound up in a good bodied wine that has kind of a creamy character to it as well, but typically has a nice crisp acidic edge too, sometimes with a zesty citrus finish.

This quaffable wine is rapidly showing up on supermarket shelves, as it (like Grigio) is an easily accessible, great chug-a-lug summer wine, that also pairs well with mild to medium-strong cheeses, and virtually all seafood. When its chilled, Torrontés also is awesome for cooling and cleansing the palate from spicy Indian, Thai and Vietnamese dishes as well.

Grüner Veltliner
The Austrian Chardonnay! Well, perhaps better described as a Chardonnay/Riesling combo for the continental county. Grüner Veltliner is the most widely planted grape variety in Austria, accounting for 40% of the country's total vineyards. Although the Czech Republic grows some too, Grüner Veltliner is the indigenous variety of Austria. Grows so easily and prolifically, that this grape was forever used as a bulk wine producer.... mostly consumed by the Austrians themselves with no great reputation or interest from the rest of the world. But lately, Austria's serious winemakers have discovered that, with lower yields and higher ripeness, Grüner Veltliner can produce stunningly intense and concentrated wines with high acidity and thus a nice crisp acid bite.

Grüner Veltliner is usually a full-bodied dry wine (up to 14% alcohol) with a firm mineral backbone, giving it the strength of character to work well with many cuisines......often surpassing even Riesling because of its ability to pair with "difficult" foods such as artichokes and asparagus. Some of these wines are now very complex, full of exotic tropical fruits, white pepper and lentils. Its kind of a blank slate to start with, and thus it can be made into a variety of styles depending on what the winemaker wants to do with it, but here are a few descriptors I have seen most associated with these wines: crisp, fruity, floral, mineral, stone, flinty, subtle peppery, herby, vegetable (like very subtle hints of asparagus or green bean).

Some are semi-sweet, most are now produced bone dry, so just be sure to ask questions about it to the retailer before you jump in and get a style that you might not want. But here is the sweet deal: most Grüner Veltliners are CHEAP! They often come in a 1000ml bottle...which is ¼ bigger than the standard 750ml bottle, but typically for about $10-$15 bucks! Hell yeah! Added bonus!

Muscadet

Muscadet is otherwise known by its technical grape name of Melon de Bourgogne, or just Melon for short. It is the most produced white wine of the entire Loire Valley region of France...dominating the most western sub-regions around Nantes, as the lovely Loire empties into the Atlantic. Its no wonder then that this varietal wine is so awesome with oysters, shellfish, or Nantais cheese...all local cuisine/wine pairings. Many Europeans consider it the single best wine for seafood pairings across the board. Some folks in Oregon and Washington are now experimenting with it too...as would make sense since they have virtually the same cool, rainy climate as coastal Loire.

To be frank, (but not Franksih! Hahahaha-pun intended!) in terms of flavor this grape is fairly undistinguished, with few strong features. Maybe that's why its so good with seafood...it doesn't get in the way and cleans the palate out nicely after each bite. But it is just so good in the role! Typical descriptors for this subtle wine include: light, tangy, hints of green apple, crisp. And Muscadet is the most common example of a wine that has flavors of **lees**....remember way back in Chapter 4: Got Wood? The *sur lies* method lets the wine sit on the dead yeasts for an extended period of time, thus picking up flavors and aromas of bread, cheese, and yeasty goodness.

It's not huge, not widely produced, not even that exciting on its own, but you should definitely try a Melon grape in its Muscadet form just so you know. Sometimes the Frenchies do get it exactly right.

Muscat

It's a single grape name, but actually Muscat is a whole damn family tree onto itself! So many different names and sub-varieties are attached to this group: you might see Moscatel, Muscat of Alexandria, Moravian Muscat, Muscat Blanc, Black Muscat, Muscato Canelli, Red Muscat, Orange Muscat, Moscato d'Asti.....the list goes on and on. Some Muscat grapes are white, others pink, still other black or red. They are grown pretty much all over the world. But there are some common family features that we can identify, and perhaps this is a wine grape that you want to tap into..... especially if you have a sweet tooth!

Because that is really what Muscats do well: make sweet, to sweeter, to excruciatingly sweet dessert wines and/or aperitifs. Muscats are extremely perfumed and floral and honeyed on the nose, with added aromas and spice according to the sub-variety. For example, I'll give you one guess as to the main aroma of the Orange Muscat wine. Yep. You got it: orange. Others have distinct peach or apricot or raisin hints. All of this family also has what we would call a really 'grapey' flavor too: what we Americans associate with grape jelly or jam or Kool-Aid. Indeed, they are so grapey that many Muscat varieties also are eaten straight up as table grapes in many parts of the world.

As for the wine, nearly every Mediterranean country has a famous wine based on Muscat, varying from light and bone dry, to low-alcohol sparkling versions, to very sweet and alcoholic potions. But the US, Australia, South Africa....hell the whole wine world... seems to have a Muscat or two that they produce nowadays. My recommendations: the perfect breakfast wine is the floral and fizzy Moscato D'Asti (the Muscat of Asti) from the Piedmont of Italy. Or Caymus Conundrum, a very popular Californian wine made from a blend of Muscat, Chardonnay, and Sauvignon Blanc. But if you like it sweet and honeyed, pretty much try a variety of Muscats from around the world until you hit the one you like best. Good luck drinking the flowery nectar!

Assyrtico , Savatiano, Vilana
I know it sounds bizarre, but I actually just want to throw these names out to you, so that you won't be scared of them when you see them pop up in the 'other white wines' section of the restaurant menu. And they look so cool too! These are Greek grapes that hardly anyone in this country has even heard of....and even fewer can pronounce. But the Greeks have been making wines since the days of Socrates and Plato and the birth of Western civilization, so we should give them some street cred. While they actually have the longest historical record of wine production, the Greek wine industry has really not even yet blossomed on the world wine stage of the modern era, for a variety of reasons that I won't get into here. The bottom line is that they are now arriving! Better late than never, and they are bringing some old-school varieties to the global market that are quite intriguing.

With the perfect climate and terrain, and millennia of wine-growing experience, the Greeks have a lot to offer in terms of great wines as their industry gets its act together. Just a few of note for now: Assyrtico is often described as Greece's best white wine grape; consider it a more minerally, steely version of Chardonnay. Savatiano is fruity, but often not exciting, wine which is often grassy, sometimes peachy or citrusy; consider it the Greek Sauvignon Blanc. And Vilana is the classic white variety from Crete with hints of citrus and tartness, with a creamy body. All go well with seafood....no doubt! They are right there on the Mediterranean, man! Back off me, Minotaur! I want to try some of these Greek wines!

Verdelho
Verdelho is, or was, a primary white grape of Portugal that really fell out of production in the last hundred years due to pest and disease. But it is making a slow and steady comeback, not just on the Iberian Peninsula (it is also grown in the Galicia region of Spain where it is called Verdello), but in Australia too! Some wines produced from Verdelho are on the semi-dry/semi-sweet side, but increasingly they are bone dry and tend to be tart and lemony, crisp and refreshing, with solid body.

Often fresh and fruity, sometimes with a honeysuckle vein, this grape can show off hints of tropical flavors such as pineapple, guava, and I see a lot of descriptions that call this the 'lime wine' because of its limey citrus edge. Foods to partner with Verdelho include seafood and light meats such as chicken, pork and veal. Spicier lemony versions go well with pesto and roasted vegetables, but if you pick up a semi-sweet one, pair it with Asian influenced dishes that have a little heat.

Significant Other Reds
Bonarda
While I have already lavished praise on the Malbecs of Argentina in a previous chapter, the grape named Bonarda is actually the most planted and produced wine in Argentina, and increasingly in Brazil. Of northern Italian origin, Bonarda has good solid fruity, jammy, plummy flavors, but not overly powerful, or over the top. And it typically has nice peppery hints that mix well with those dark fruit flavors. I know this is a stretch, but I consider Bonarda a kind of Latin American cross between Merlot and a light Pinot Noir. Its got light fruitiness and just kisses of earthy spice that together make it intriguing.

Wines made with the Bonarda grape are immediately drinkable and known for the light body and fruitiness. With a soft tannic component, Bonarda pairs well with roast beef, salmon and most soft cheeses. It's a great, easy picnic wine. And inexpensive! You can even grab reserve Bonardas that have a decent oak spine for less than 15 bucks! Stellar red alternative!

Nero d'Avola
But speaking of stellar red alternatives, I now have to show my cards and tell you about one of my personal favorites, and a well kept secret up to now: The Nero! The "Black Grape' of Avola is one of the definite rising red stars in the world, but has been little known outside of its native home in

Sicily up to this point. That's going to be changing soon. This succulent Sicilian is typically rich and perfumed, with ripe dark berry aromas like plum and cassis, with a touch of raisins, leather, and sometimes even almond thrown in. Wow!

The fruitiness is very forward in inexpensive bottles, but Nero d'Avola also has awesome oaking potential, and a lot of Sicilian winemaker's are just now starting to dabble with pushing it to the limits. When you get a reserve Nero that has had some time in oak, you will be floored on how complex and integrated the flavors are, but it will still be completely drinkable immediately, even when young. I'm sure that some of them have aging potential....but I always end up drinking mine up as soon as I get them!

Given its intense fruit, and the way that it easily incorporates the wood flavors of oak, I fully expect this unique little varietal to be in the kings of world wine category in the next decade or two.
But no one knows about them right now! So they are an exceptional value! Get them while they are cheap! I have drank cases and cases of an $8 Nero that was a solid table wine, but was even more phenomenal at that low price!

This is what really excites me about today's wine world too: the fact that this unknown wine is finally being put forward as a unique grape with unique expression from its unique terroir in Sicily....and it's great! How many more 'local' varieties worldwide are still unknown to us? People! Listen to me! Don't waste your whole life trying mediocre Merlots from every corner of the planet! Try the local wines that are specific to distinct regions!

Petite Verdot
Everybody back the hell up! This grape is not taking any palate hostages! Petite Verdot is one of the traditional classic grape varieties that go into Bordeaux blends...and typically in very, very small amounts. Rarely made into a varietal wine....and for good reason!
This grape is sometimes referred to as the winemaker's 'spice box' because you just use a dash of it to spice up your wine—its dense fruit, black color, heavy-ass tannins and powerful flavors pack a serious wallop! Its got a little of all this: vinous, black fruits, blackberry, pencil shavings, molasses, tar, weeds, nettles, leather, cedar, and cigar box. And intense! A little goes a long way, and when used too heavily in blends, can make the wine coarse and unrefined.

So why would I suggest it is a significant other red grape to know? Because bigger is better in America, baby! While the Europeans don't grow or use much of the stuff, it has proliferated in California and Australia, where prime growing conditions have made it even more over the top! Given the current 'New World' wine rage to push wines to the extreme, I have a feeling you are going to see more and more of these produced as consumers look for the next big rush. Its kind of like the bungee-jumping of the wine world: an extreme experience that thrill-seekers will seek out.

Carmenère

The 'Merlot of Latin America', Carmenère was transplanted from its native Medoc region to Chile and Argentina back in the 1850's....and much like Malbec and Bonarda, has had much more success as a varietal wine there than it ever did back in Europe. Extremely similar to the Merlot, but with a bit more spice and earth characteristics. So similar, in fact, that massive planting of Carmenère in Chile were mislabeled as Merlot for a hundred years. It gets good color, good dark fruit flavors, but with an added complexity of herbal to gamey aromas...almost reminiscent of a weaker Cabernet Franc.

I won't lie: I don't think this variety will ever achieve any level of greatness, even when created by a master winemaker. But, it does make for an interesting South American red that can be acquired for not many pesos. A solid, inexpensive table wine that is great for college students and others on a tight budget.

Dolcetto

Hey there little buddy! Don't keep hiding in the shadows! This poor wine grape is still under the radar   for most of us, because it is so wildly overshadowed by its big brothers in the northern Piedmont: the big tannic Barolos and Barbarescos. Hell, I even ignored it in the Italian Stallion chapter in this very book! Roughly translated, Dolcetto means "little sweet one," but these wines ain't exactly sweet...in fact, not at all; they are more often described as light and fruity. The Italians use this stuff for everyday drinking wines, and it is wonderful for that.

Dolcetto wines that are soft and fruity and usually ready-to-drink when released, although now some folks are crafting them for extended aging too. Nose: red flowers, lilacs, tulips dried cranberry, fruity, jammy, cherry. Mouth: fresh fruit, cherry strawberry, blackberry, jam , little leather, barnyard. The Italians like this wine for everyday drinking because of its soft tannins, ripe fruit, and ability to match with a variety of foods.

Awesomely priced too: for $10 to $15 bucks, most people drink mediocre to sucky Sangioveses or Chiantis, instead of drinking these solid Dolcettos...which pair beautifully with all the same Italian foods that the big boys are good at: spaghetti, and pasta, and red sauces all the way around! Save a buck, and try this other Italian grape that has just as much play!

Touriga Nacional

Portugal's finest! What Tempranillo is to Spain, what Sangiovese is to Tuscany, what Cabernet Sauvignon is to Bordeaux....such is Touriga Nacional to Portugal. It is the king, the backbone and the bomb to the Portuguese wine industry. This is the grape that plays a big part in the blends used for the best port wines of Portugal....but the reason it is included in this 'significant others' section is because it is increasingly being used to make varietal table wine of some significance. And not just in its home state: the US, South Africa, South America, and Australia are experimenting with it too.

So why the hype, and he likelihood that you will start seeing these wines in increasing numbers on the shelves? It is a beast! Big huge perfumed nose, this thing reeks of violet, cherry, raspberry, blue and black fruits of all kinds, allspice, and even hints of cocoa are common. Exploding in the mouth with over-ripe berry flavors, pepper, cherry, leather....and with huge tannic structure that makes for bold, big muscular, rich mouthfeel and texture. Insanity! I believe you will see some magnificent versions of these wines, at decent prices, and they will possess unlimited aging potential too. Get in on the experimentation action of Touriga Nacional while it's still kind of new, and exciting!

Agiorgítiko and Xynomavro

Agiorgitiko and Xynomavro are the two red grapes/wines which will make the biggest splash on the world stage for the up-and-coming Greek wine industry. Again, I'm not trying to promote the Greeks because I'm getting some sort of kick-back or anything, but they just have so many unique grapes that no one has even heard of yet! And many of them are really solid! Plus, I had to have at least one word in the book that began with the letter X!

Just know this for now: Agiorgítiko can be considered the 'Greek Merlot' but with a bit more sass and spice. On the other hand, Xynomavro is basically the Greek words for acid and black, which hints what this grape is all about. It is workhorse red grape of Greece, producing a wide variety of wines ranging from chewy, tannic reds to fruity rosés. Sometimes compared to Nebbiolo, other-times to Cabernet Sauvignon, it has deep color and high tannins and bright acidity. Big wine boys of the future...Zeus would be proud!

So much for our significant others...

To restate, there are many, many other great grape varieties not touched upon in this chapter, nor the book as a whole. I'm just pointing out the big boys you must know, and then these lesser known varieties that are made into single variety wines that you will see increasingly in the American market. Lots of other grapes make lots of other great wines....but are typically used as a component of a blend, and therefore it is impossible for you to try them on their own to discover what they are about. But never be afraid to experiment and try a wine that you've never heard of before... discovery is half the fun!

Lesson 19: Before & After the Meal: Apéritif and Digestif Wines

What are aperitifs and digestifs. What is the difference between table, fortified, and sparkling wines. How these wines are produced. A wide variety of examples and styles from each of these wine categories that are commonly used before and after the meal.

Peoples! We American drinkers need to get our act together and get classed up, and return to the good old habits that the Europeans themselves have started to slack on...and that involves well-rounded wine consumption not just with the meal, but before and after the meal too!

Pre-gaming and post-gaming wine drinking? Absolutely my friends! Of course, there are lots of great table wines which we have talked about in chapters past that would work just fine as bookends to the meal itself, but I wanted to use this chapter to introduce you to some of the 'oddballs' of the wine world, many of which are crafted specifically for this job. Port and Sherry are the sentinel classics of the 'other wines' world, but there are tons of other unique tipples to consider for the dinner party too!

Let's get technical for a second: An **apéritif** is an alcoholic drink that is usually served to stimulate the appetite before a meal. I can attest to the fact that they really work well too! A warming alcoholic tipple gets the stomach juices flowing instantaneously, and makes the oncoming meal even more desirable. On the opposite side of the dinner party is the **digestifs**, which are served after meals either as a compliment to dessert.... or as a dessert in their own right. Sherry is a very popular apéritif from Spain. In Greece, Ouzo is a popular choice; in France, Pastis; in Italy, Vermouth or bitters. And the British would never be satisfied if a meal was not concluded by downing a whole bottle of Port. The apéritif is the warm-up; the digestif, the cool-down dessert. And you thought drinking was just for bars! Pshaw, pshaw you commoners! We have much work to do here!

Now, there is no single alcoholic drink that is always used as apéritifs and digestifs; liquor-based cocktails and liqueurs are common choices in our culture. However, there is a great variety of wines and wine-based beverages that are even better...and of course are more civilized than tequila shots prior to your entree arriving; I guess that works fine for your Ruby Tuesday experiences, but why would you ever want to repeat that shame? You drink wine now, so its time to step it up across the board, and

at least be able to reference the other major wine categories. That is the focus of this chapter: fortified wines like Port or Sherry, sweet/dessert wines like Sauternes or Icewein, and sparkling wines like Champagne or Prosecco. Let the pre-game begin!

Like the last chapter, there is no

specific tasting lesson attached to all these wines. I simply wanted to provide you with a quick read for reference, and to hopefully pique your curiosity enough to give some of these wines a try. This is a wildly generalized overview of these wines, and much more detail exists on each of the types I will talk about here, but I felt compelled to at least introduce them to you. Students always ask me tons of questions about these wines, although they are nowhere near as widely consumed as standard table wines. Which brings me to the first...

Fortify your Sparkling Table...or is it Table your Sparkling Fortification?

We have to throw just a few more primer terms out here first to make sure we are all on the same page. A **table wine** is a generic term to most of the wine world which alludes to what (for a lack of a better description) I will simply call 'regular wine'. It is fermented grape juice that usually has anywhere from 8% to 16% alcohol content, and it is the stuff we have been talking about for this entire book! California Chardonnay, Bordeaux, Chianti, Pinot Noir from Burgundy...these are all table wines.

Now, I said table wine was a generic term to most of the world, and I stick by that statement, but you should be aware that many European countries have a strict legal definition of what a table wine is, and it is tied to level of quality of the wine. And in the US, our strict legal definition of table wines is tied to alcohol content, which is supposed to be less than 14% for these wines. For purposes of this book and the rest of your life, you need not really give a rats ass about any of the 'strict legal definitions' from either side of the Atlantic. Both are increasingly meaningless in today's world. Table wine, is 'still' wine we drink with meals...at the table!

But why did I just say still wines with 'still' in quotes? Because that brings us to the second major type of wine: sparkling. **Sparkling wines** are wines with bubbles! They are carbonated, just like soda pop! Since they are charged with carbon dioxide, they are under pressure, and the top will 'pop' when opened, and then bubbles of the gas will then fizz out of the liquid. It is fizzing, and active, and moving....the opposite of 'still' table wines. Get it? Good. I'm sure you are familiar with these too...at least the most famous version we call Champagne, but there are several others to consider. Let's beef it up for the last category:

Fortified wines are those wines to which a neutral distilled spirit (liquor, that is) has been added to bump up the alcohol content to the neighborhood of 20% by volume or higher. These higher alcohol wines are typically—but not always— high in sugars too which balances out the body and flavors of the beverage...go back and read over Chapter 7: Rock Your Body but Keep Your Balance if you have already forgotten this stuff. There are lots of different ways to increase the sugar in the wine and/or concentrate the alcohol to higher levels, which gives us a wide variety of wines of radically unique styles within this category. Many fortified wines are employed as apéritifs and digestifs...and even for dessert itself which brings us to...

Dessert wines as a general rule are always sweet, and are paired with the dessert course or just consumed as the dessert itself. The reason that this definition is intentionally vague is because some really sweet *table* wines could be classified as dessert wines, many *fortified* wines are used as dessert wines, and even some sweet *sparkling* wines are used as dessert wines. So it is a more nebulous, generic classification that pulls examples from all three of the major types of wine, but I think it will help you understand the bigger picture of what all these wines are about.

Let's start with the sweet table stuff and work our way forward for our rundown reference of oddball wines used for before and after the meal.

Sock it to my Sweet-tooth: Table Wines for Dessert
There are many ways a winemaker can increase the sugar levels of a standard table wine to get it to the saturated state of being a dessert in its own right. Back in Chapter 5: Dry vs. Sweet, we talked a little about directly adding sugar to the grape juice (chaptalization) and the process of holding back unfermented grape juice and then adding it to the finished wine in order to increase sweetness (süssreserve). Both processes can up the sugar/sweetness of the finished wine. For this chapter, I just want to cover a couple of other ways to achieve these ends: 1)just grow a really sweet grape, or 2)remove water from the grape in order to concentrate the sugar that is already there. Ah! That second option is the most intriguing and interesting and involved...so let's handle the easy one first:

Grow a really sweet grape!
Not all wine grapes are created equal, and some grape varieties just naturally have really high amounts of sugar when ripened fully in a good season...and the winemaker can use some of that sugar to produce the alcohol, but have enough sugar left over to make the wine taste sweet. Sweet! On top of that, the winegrower can intentionally leave the grapes on the vine past their average ripeness stage if the weather is good. Letting the fruit hang for even longer makes it 'over-ripe' as the grapes start to shrivel, and maybe even rot a little...kind of like leaving that banana on the countertop too long—you can really smell the pungency and the fruit gets sickeningly sweet. In the vineyard, the grapes become concentrated in flavors and sugars are maximized. If the grape variety itself also has honeyed, grapey, floral aromas and flavors, then it is a prime candidate to make into really sweet and unctuous dessert wine.

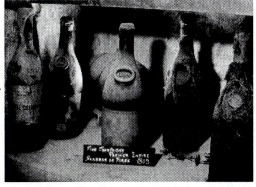

Grape varieties/wines that fall into this category are Huxelrebe, Riesling, and virtually all of the Muscat family. While dessert wines made from Huxelrebe and Riesling are mostly a German specialty, you can increasingly see other wine regions experimenting with them too. Muscats on the other hand are a tremendously diverse 'family' of grapes which can be found all over the planet: You will see reference to Muscat of Alexandria, Orange Muscat, Black Muscat, Moscato d'Asti, and dozens more examples. Muscats are extremely perfumed and floral and honeyed on the nose, with added aromas and spice according to the sub-variety; Huxelrebe and Riesling are too, but not as pronounced, which actually makes them more accessible to most palates. And the German Rieslings and Huxelrebes that you want for dessert will be labeled as Auslese and higher....oh crap, I haven't even told you about that stuff yet, have I?

Well before I go any further, I might as well deal with the Germans for a minute...just as the British should have prior to World War Two. No, not those Germans! I mean German wines. Because a whole slew of German wines fall into the dessert wine category, and their whole system of wine labeling is based on sweetness levels, so read the inset box on the opposite page. Do it now! Schnell! Schnell!

Are you back? Good. So Muscats the world over are typically semi-sweet to very sweet table wines that make a delightful dessert; German Hexelrebes and Rieslings do too....especially the ones labeled as Spatlese and Auslese. These are grapes that naturally possess floral, honeyed aromas and flavors coupled with high sugar levels for alcohol production, with enough to spare for sugared sweetness. If you just finished reading that inset box on German wines, then I've already let the cat out of the bag on the next big sub-category on how to make sweet table wines, and that is....

Dehydrate the grape!
To remove water from the grape in order to concentrate the sugar that is already there...and there are several, often bizarre and funky, ways to achieve this. We can rot the grape, we can dry the grape, or we can freeze the grape. All eliminate water and focus the sugars, and the flavors, of the grape. Most (but not all) of these processes are employed to make dessert wine. Let's get back to the Germans for starters and expand from there...

•*Rot the grape!* In cooler, damper climates winemakers can actually intentionally employ a fungal infection to dehydrate the grape. While there are a plethora of horrifically bad fungi which attack grapevines, there is one fungus among us that we encourage: *Botrytis cinerea*, otherwise known as 'noble rot.' Why so noble? As this fungus attacks the grapes, it coats the skin and causes the water inside to evaporate. With the absence of water, the sugar becomes more concentrated and the botrytis actually begins to positively alter the acidity within the grape, making for an added dimension of complexity, richness, and flavor that can only be achieved by this rot. It sounds gross, but it's delicious!

It's a tricky process, and does not happen every year; usually only in cooler

Most all of the German wine you will see in the US are from the top level of their classification system: Pr√§dikatswein or Qualit√§tswein mit Pr√§dikat (QmP)... roughly translated to 'quality wine with distinction.' Like many other places, Germans usually label the wine by the grape inside it (like Riesling or Huxlerebe) and also by the region where it is produced. *Unlike* any other country on earth, Germany also labels wines according to the the sugar content of the grapes/juice at harvest, and these terms are strictly, legally defined. Levels of grape sweetness are so important to this system, that these designations have become a rough indicator of quality...and price. All QmP wines will have one of these words somewhere on the bottle, and it is telling you something about the wine inside.

In order of increasing sugar levels in the must (grape juice):
- **Kabinett** = regular harvest, fully ripened grapes. Typically makes semi-sweet light wines, with crisp acidity, but can be dry if designated so. An American palate would say this wine is "a little sweet."
- **Spätlese** = "late harvest" They let the grapes hang out even longer on the vine, as described earlier. Typically semi-sweet, often sweeter and fruitier than Kabinett. An American palate would say this wine is "sweet."
- **Auslese** = "select harvest" Made from selected very over-ripe bunches of grapes, semi-sweet to sweet, sometimes with some noble rot character. They actually go thru the vineyard multiple times over several days individually picking out the over-ripest bunches of grapes, and letting the others hang even longer. An American palate would say this wine is "really sweet."
- **Beerenauslese** = "select berry harvest" Made from individually selected overripe grapes often affected by noble rot, making rich sweet dessert wine. Dig this: they go thru the vineyard multiple times over multiple days selecting out the over-ripest rotten individual berries! Talk about labor-intensive! An American palate would say this wine is "wow! overwhelmingly sweet."
- **Trockenbeerenauslese** = "select dry berry harvest" Same as Beerenauslese, but the shriveled, noble rotting berries are now turning to raisins on the vine, concentrating sugars even more. An average American palate would simply explode when subjected to these extremely rich, sweet wines.
- **Eiswein** = ice wine They leave the rotting, drying shriveling grapes on the vine until the first really hard deep freeze. Then they pick the forzen berries on the vine and ress out the juice as quickly as possible, making a very rich, concentrated wine.

By the way, any of these harvest levels of grapes could be made into a semi-sweet or a bone-dry wine if the winemaker wants to let the fermentation go all the way. It does not happen a lot, but if this is the case, you will see the words 'trocken' (dry) or halb-trocken (half-dry) right there on the label to alert you to the style. Otherwise, the stuff is sweet! General rule: all German wines labeled Auslese and higher are easily used as dessert wines. Now get back to the story to find out more about that noble rot and ice wine stuff...

climates when the dampness/dryness conditions are exactly right. And as you can imagine, dehydrating your grapes means there is a lot less juice let to make wine out of, so you have less wine! Couple that with the intense labor associated with making these wines, and you may now be starting to understand why the prices can get really outrageous, really fast! But these dessert wines have huge body and mouthfeel, volumes of complexity, and intense honeyed, raisiny, floral, citrus, peachy, tropical fruit, mineral and even diesel (yes! like the fuel!) aromas and flavors that are awesomely integrated into a single beverage. A few notable 'rot' wines of note:

❖ *Germany:* As alluded to above, the German designations of **Beerenauslese** and **Trockenbeerenauslese** both employ the rot, and the classic Riesling grape is often the gold standard for these types of dessert wines. However, you can now find Hexelrebes, Silvaners, and even Gewürztraminers in these dessert styles too.

❖ *France:* **Sauternes** dessert wine is from the Sauternais region of the Graves section in Bordeaux. Crafted from a blend of Sémillon, Sauvignon Blanc, and Muscadelle grapes that have all undergone the rot, this stuff is liquid gold...and priced accordingly.

❖ *Hungary:* Consumed by royalty of central Europe for centuries, Aszú-style Tokaji or **Tokay** wines are from the Tokaj-Hegyalja region and are a blend of the Furmin grape, and a handful of others you have never heard of, nor will ever hear of again. But they do undergo the rot, and make magnificent amber-hued nectar wines! The cellars where the wines are aged are even covered in fungus! I think Franz Ferdinand is still hidden under a case or two of Tokays in the imperial cellars! Drag his old dry bones out of there! And speaking of drying...

• *Dry the grape!* You know...make raisins! Well, maybe not all the way to raisin level, but close. While noble rot is in effect drying out the grape, this practice alludes to just air drying the grapes...which is the raisin making process, no fungus needed. Fungus only works in cool damp climates anyway, so air drying is better for hot, dry climates like the Mediterranean region.

Now, raisin wine has been made since ancient times, but we need not concern ourselves with Caesar or his salad for right now. However, Italy is still home to a number of 'passito' or 'straw' wines....so named for the straw mats that the grapes are laid upon and then baked in the sun in order to be dried out in the classic method. Today, the clusters of grapes are more often hung on wooden racks, and sometimes on rafters in attic spaces where high drying heat is achieved. After the wine is produced, they often see extended aging in small wooden barrels, under similarly hot conditions.

The dehydration effect has similar outcomes, but with different flavors: concentrating sugars and flavors across the spectrum of dried figs, raisins, dates, melon, tropical fruits, candied apple, yellow plum, and almond. The wines are 15% alcohol and up, typically sweet to very sweet white wines (although some exceptions are listed below), similar in density and sweetness to Sauternes and capable of long life. Here are several 'straw' wines of note, mostly of Italian origin:

- ❖ **Vin Santo**: This traditional Tuscan wine is made from Trebbiano and Malvasia grapes, and develops a deep golden or amber color, and a sweet, often nutty, taste. Awesome with almond biscotti.
- ❖ **Sciachetrà** is the name of a similar passito whie from the Cinque Terre region on the NW Italian coast—it is a five-town inhabited national park that is quite literally one of the most beautiful places on the planet, and their dessert wine adds to this image. Que bella!
- ❖ **Recioto di Soave** is the passito white wine from around Verona, made from the Garganega grape used in Soave which is an important wine region in northeast Italy.

Speaking of Verona...just to the east in the region named Valpolicella, they also do the dried thing for some red grapes too which are then fermented in two different ways to make both a sweet red and a dry red wine:

- ❖ **Amarone** or **Amarone della Valpolicella** is perhaps the most famous of these passitos, although it is completely dry, and therefore not a dessert wine choice....but sooner or later you will encounter one because they are quite famous, so what the hell, I should throw them in here. A big, bold, very ripe, raisiny red wine with very little acid and usually in excess of 15% alcohol by volume; it is much like a big California Zinfandel, but with very unique flavors and aromas added in by the drying process itself. After raisinating the grapes, the wine production is allowed to go to completion, thus making it both dry in style and big in alcohol. Amarones are then often barrel-aged for 4 or 5 more years, making them even more complex and rich. However, if the fermentation process gets stuck, stops, or is intentionally stopped, there will be residual sugar left in the wine making...
- ❖ **Recioto della Valpolicella**, a sweet version of the Amarone described above. Both concoctions are crafted from the Corvina, Rondinella and Molinara grapes, which you really won't find outside their NE Italian homeland. The low yields and labour-intensive production methods employed to make all these straw wines equates to expensive prices, no matter who the producer or how good/bad the vintage.

But the Italians don't have the entire dried-grape-dessert wine category completely cornered! **Prošek** is a similar wine from the Balkan region of Europe, but man, that stuff is expensive! California and Australian producers are increasingly experimenting with the style too, so you can find New World representations on your wine store shelves. Any New World label which makes reference to 'recioto,' 'passito,' or 'straw wine' is probably made in this style. And of course the Frenchies are huge into all things alcoholic, so I just want to throw in this single term which they use:

Vin de Paille is the French term for these dried-grape wines, and are made from various grapes in various regions....like Jura which makes a Chardonnay based blend, or Alsace which makes a Riesling-dominated one. Most have flavors of peaches and apricots and are excellent with foie gras. But we need to get on with this section! Let's wrap it up already! The last way we can concentrate sugars is to....

•*Freeze the grape!* And we have already alluded to this process back in that inset box on German wines. Given its northern latitudes and much cooler climate than any other wine-growing region in the planet, Germany is the inventor and still foremost producers of these types of wines. The end result: **Ice-wine** or **Eiswein**. How do they do it?

A winemaker can elect to roll the dice, let the grapes start the noble rot process, which turns grapes to rotten raisins, and then gamble even further by letting the grapes continue to hang until the first big hard freeze occurs. The freezing process locks up all the water in the grape as ice, and they rapidly harvest them (usually at like 3-5AM when temperatures are at the lowest), get them to the winery and press out the remaining ultra-concentrated juice....which without the water has the consistency of maple syrup. This syrup is super-high grape juice sugar concentrate, and produces exceptionally rich and complex dessert bombs! Sweet! Literally!

With the advent of refrigeration and freezing technologies, any winemaker from anywhere on the planet can now use machinery to make an Ice-wine, simply by freezing their grapes. Don't expect those 'false freeze' versions from warmer climates to have the depth and complexity of the real deals. However, some truly cooler climates outside Deutschland are doing a dandy job of it too: the Canadians have a lot of solid Ice-wines to offer nowadays, usually for a good price too. Typical grapes most used for the icing process are: Riesling (which predominates the German versions), Huxelrebe, Vidal (most used in British Columbia and Ontario) and a seemingly bizarre pick which is actually quite delightful and growing in popularity: Cabernet Franc. Cool! Literally, cool! Its Ice-wine!

But enough of the coolness...let's bring the heat now! What can really get you heated up? Why, more alcohol of course! We must press on and be brave enough to handle higher amounts of alcohol now. But how to instill ourselves with this fortitude? Why, by consuming fortified wines of course! Fortification for your soul, with extra spirits for your spirit! Ha!

Let's beef it up! Fortify my aperitifs and digestifs

You've now read about a variety of ways to modify and concentrate the sugars in the *grape* in order to produce a wide variety of sweet wines. Up to this point, I have only talked about sweet table wines—but now it is time to get to another classification of wine altogether. Most (but not all) of the wines I will talk about for this section are also sweet, but they have been manipulated during or after the *wine*-making stage to create/modify their sweetness levels. And this manipulation centers on **fortification** or

'mutage'.

See, some winemakers don't want to wait around for sugars to be concentrated naturally by raisins or ice or fungus, so they take matters into their own hands. Fortified wines are a whole distinct class of wines, defined by the fact that they have been beefed up alcohol-wise to a higher-than-natural proof. They contain characteristics of table wine and the grapes that make them, but have that added alcohol punch which sets them quite apart from the norm...and makes them warm the insides more like a liquor or liqueur. What a combo! Wine flavor and complexity with a slight liquor kick! There are some dry versions that are totally excellent aperitifs, but the bulk of these bad boys are bodacious big, sweet, stand-alone desserts in their own right....and all are excellent for swirling and sniffing in a big snifter while wearing your smoking jacket in front of a roaring fire.

So how, and when, would a winemaker fortify a standard table wine? The *how* is easy. At any given point in the process of making wine (from fermentation to blending to aging to bottling) the winemaker can dump a hell-ton of raw spirits, typically brandy, into the batch. Brandy itself is simply distilled wine: heating wine until the water boils off, which concentrates the pure alcohol to around 70 or 80% by volume. You can go to the liquor store and get all sorts of finely crafted and fancy brandies, like Cognac, or nice brandies that have been infused with herbs and spices, like B&B. But for our story we are talking about bulk brandy of no particular pedigree which is desirable just for the alcohol itself.

Now, *when* exactly the winemaker dumps this high-test brandy into his batch of wine makes all the difference. If brandy is added *during* the active fermentation process (when yeast are converting sugar into alcohol), then the high % of alcohol shocks and awes the yeast, thus ending the fermentation process. When this happens, there is still lots of sugar left in the wine that the yeasts had not yet gotten to before their untimely death....thus the wine will taste sweet. If the winemaker allows the fermentation process to fully finish prior to the additional of brandy, then all the sugar has already been consumed by the yeast, and the resulting wine would be dry. In both cases, the wine has additional alcohol blended in, bringing the beefed up total to anywhere from 18 to 23% (standard table wines range from 10 to 15%.) Game on!

You are likely already familiar with the two most popular examples of fortified wines: **Port** and **Sherry**, but there are a variety of lesser known fortified wines of note as well. Let's dabble a little with all of them, and we will start with the big boys first...

Port
Port wines are the premier product of the Douro Valley in northeastern Portugal, named for the major coastal town which serves as the export and cultural center of the style: Oporto. The style of most Port is distinct:

powerful red wines with high alcohol level (20% and up), high sugar/ sweetness level, and usually a ton of body and flavor to boot. Just the right amount of alcohol to give a slight burn, but with a smoothing sweetness to smother the fire. It's a winter warming mouthful! The high sugar content is due to the fact that the winemaker dumps in that neutral brandy well before the fermentation is finished, thus stopping the yeast in their tracks and maintaining a lot of residual sugar in the wine.

These wines have been so mass produced, so rabidly consumed (mostly by our drunken British brethren) and so stylistically successful that the term 'port' itself has become a generic word for any fortified wine of this style. So you can easily find 'port' wines from California, Australia, or just about anywhere anymore. Most wine-makers will tip their hat to respect and tradition, and will label their wines as 'port-style' instead of calling it port. Most, but not all. However, true Port is only from Oporto.

And it is a blend of 5 grape varieties: Touriga Nacional, Tinto Cão, Tinta Barroca, Tinta Roriz, and Touriga Francesa. Rarely are any other grapes used, and rarely will you see these grapes outside of Portugal… with the possible exception of Touriga Nacional which is the true backbone of the Port blend, and a potential powerhouse of varietal wines in the future. These sweet red wines are often served with dessert, or as dessert alone. The high sweetness level is actually offset on the palette by the high alcohol, which accounts for its great balance and also allows the flavor of the wine itself to be exposed. Just a few sub-styles for you to be aware of: *Ruby, Tawny, Vintage* & even *White*.

- ❖ **Ruby Ports** are your entry-level port for the novice…the 'training wheels' you will need before you try the big boys. More simple, fruity, and slightly lighter, these ports are made with the classic 5 grapes, and are aged for two years in oak tanks before going straight to the bottle. Thus, they retain a lot of full, sweet fruit character (fig, date, licorice, cherry, raspberry) with some hints of wood, and are not terribly complex….nor very expensive! Not built for age, drink them young. Give them a go for starters!
- ❖ **Tawny Ports** are the heart and soul of most port production. Wines blended from various harvests of grapes, and usually from various vineyards, they are blended together and then aged in oak casks anywhere from 7 to 40 years before bottling. This slow, long, wood aging process is also slowly intermixing oxygen with the wine… but in this high-alcohol, high sugar atmosphere, the destructive bacteria which ruin most wines (when oxygen is introduced) cannot survive. Thus, they cannot 'go bad' on your counter-top after you open them. After bottling, the stuff is good to go, ready to drink, and will have much less fruity, and much more 'woody' and oxidized flavors infused within it: nutty, toffee, coffee, walnut, butterscotch, caramel flavors come to mind, with only a hint of grapiness.
- ❖ **Vintage Ports** are just the opposite: crafted from a single harvest, and sometimes even a single vineyard, these big red wines are

bottled just after a couple of years in barrels (4 to 6 years for some types) and are extremely expressive of the fruit itself, containing a big jammy, fig, date, licorice, blackberry, cassis, and raspberry blast. As Tawny Ports are 'wood aged,' the Vintage Ports are 'bottle-aged.' These are the ones that age in the bottle for decades, and can become quite expensive when you start dabbling with the higher quality houses with the big reputations.

❖ **White Ports** are also appearing more often on the shelves nowadays too. Made from white-skinned grapes (thus, not the big Port 5), these ports are white wines that are still go through the fortification process and thus have high levels of alcohol. However, they are made in a wide range of styles, from all the way dry, nutty wines which are great as aperitifs; to all the way-sweet wines more typical of port-style, which are great for dessert. These white ports have great diversity very similar to Sherry wines, which we will get to in a minute.

Which brings us to the last..so which ones to buy? Unlike most other wines, the grapes that go into Port production are not of great consequence, and quite honestly there is not huge variation from year to year; it's really all about the producer. The house name that blends and bottles the stuff carries the reputation of its style and quality. Look For anything from Dow's, Warre's, Rameriz, Fonseca, Sandeman, Harveys of Bristol, or W. & J. Graham. All of their stuff is stellar. Don't want to spend that much? Then check out any 'port-style' wine from another country that's not Portugal. They will do in a pinch, and can often be quite inexpensive. I would recommend Hardy's Whiskers Blake from 'down under' as a good starter port. Oh! If you are a chocolate fan, grab a bottle of Trentadue's 'Chocolate Amore' port from California; the name says it all. Which reminds me, you simply must try any Port wine with some chocolate. Dark chocolate, rich chocolate, any chocolate! Wow. Vintage Port with blue cheese or any other intense, hard, aged cheese works well too.

So grab yourself a big-ass ol' brandy snifter, and pour yourself some Port... its flavorful fun and warming wonderfulness will whisk away your winter weariness. That soulful burn will get you to the Spring for sure! By why wait for seasonal change? Let's skip to the next fortified friend which has a style for every season...

Sherry
The next, and perhaps most diverse, fortified wine is Sherry. Made almost entirely from the Palomino grape grown in and around the southern Spanish city of Jerez, Sherry is a wine/style that has been around at least since Shakespeare's time (he alludes to Sherry in many of his works) and is currently one of the most under-appreciated, underdog wines on the planet. Why under-valued? I guess it has just

gotten a bad rap as being an over-sweet and over-nutty wine, which some sherries are, and that is just not in vogue in todays/ wine world. But as alluded to above in the White Port section, sherries come in a huge range of style, sweetness and strength, so ask you local wine rep exactly what you are getting into before purchasing one.

However, all sherries have at high levels of alcohol (16%-18% on average) and at least a little of the awesome rich nutty flavor that make them distinct....so how do they get that high alcohol and unique flavors? In a very interesting process that involves more fungus. No, not the noble rot! That is for grapes! What we are talking about for sherries is modifying the wine with a floating yeast called *flor*. The floating *flor*? Too fabulous!

Let's have some fun with flor: Sherry production starts pretty simply, in the same way all other table wines are crafted. Using the blandly-average-to-taste but highly acidic Palomino grape, the sherry maker produces a standard white wine. The neutrality of the Palomino wine is actually a great feature, because it provides a blank canvas for the winemaker to start with. At this point the neutral brandy spirits are put in—unlike port wine which fortifies during the fermentation process, sherry wines are fermented until completely dry and then fortified. So sweet sherries must get their sugars from someplace else... hmmmm...but I am getting ahead of the story.

Next, the wine is put into barrels, and this is where the flor comes in...this floating yeast forms a film similar to pond scum across the top of the liquid inside the barrel. The flor provides a semi-permeable barrier between the air and the wine, so rapid oxidation does not occur: for some sherry types, no oxygen at all gets in; for other types the oxidation is slow and controlled so alcohol-destroying bacteria are held at bay. It is this slow interaction with air, wood, wine and mostly the flor that imparts these unique nutty, tangy, yeasty, citrus aromas and flavors. Some sherries will form a thick layer of flor, others just a little, thin layer, and some no layer at all...into which more distilled spirits may be added to ward off bacteria and flor alike.

Bored of flor yet? Yeah, I know, this wine drinking stuff is not supposed to be this tedious and fact-packed. But you have to know this stuff in order to get the sherry that is right for you! All sherries start out in the ways described above, but then go through various aging, blending, and barreling movements which I will not bore you with here. (Want more info? Goggle: the Solera system) If you have followed me so far, then you are ready to understand this abbreviated list of sherry styles, because they are radically different from each other and diverse enough to be used before, during and after the meal! Not many wine categories can make that claim! So let's do the Sherry Stylistic Shake-down:

❖ **Fino** is the fully dry and totally pale sherry that had a thick, chunky flor cap...which means no oxidation, and lots of nutty tangy flavor. Absolutely fantastic aperitif. A chilled Fino with almonds, olives and even shellfish is as close to heaven on earth as you will get in Spain. You might also see bottles labeled **Manzanilla**, which is a variety of fino sherry produced specifically around the port of Sanlúcar de Barrameda, and it has additional salty characteristics.

❖ **Oloroso** refers to a sherry that had a very thin or weak layer of flor, and therefore was subject to some oxidation which brings about its deeper dark color, oak flavors, and richer style. To prevent spoilage during its aging, it is bumped up even higher with spirits, making it the most potent of the sherry styles. It is the main base for other styles listed below.

❖ **Palo Cortado** is one that you won't likely come across, but I will give it a shout-out anyway. It is a rare breed that initially aged under decent flor like a Fino but, due to a variety of factors, it then oxidizes and develops a character similar to Oloroso. So its got the color and body of an Oloroso with the nutty crisp hints of a Fino.

Those are the three sherry 'building blocks.' They are bottled under those stylistic names, and now you know what they are all about. But the sherry-maker can also take those wines still in the barrel, and manipulate them further to create even more styles:

❖ **Amontillado** is a Fino sherry that has been aged first under a cap of flor and then later during an extended aging period is slowly exposed to oxygen, which makes it darker than fino but still lighter than oloroso, maintaining its nutty flavors and adding oak components. Real ones are old and rare and highly sought after.... ever read Edgar Allen Poe's *The Cask of Amontillado*? Scary wine story!

❖ **Pale Cream Sherry**: start with a fino sherry base and add a sweet white wine. Thus it has a light color, but with a sweet finish. We will get to the sweetening agents in a minute.

❖ **Cream Sherry** is a blend of oxidized olorosos and sweet white wine. Thus, is darker in color, richer in flavor with sweeter finish. Maybe you have heard of the most famous brand, Harvey's Bristol Cream?

❖ Full on **Sweet Sherry**: this is more of a generic category that would include the aforementioned Pale Cream and Cream sherries, but would also include the top of the sweet heap: **Pedro Ximénez** sometimes just labeled as **PX**. Pedro is actually a type of grape that makes an intense dessert wine on its own accord, but you will often see many sherry makers use it to make intense sweet sherries. You may see the name of the grape right on the bottle.

So you see that the really sweet versions of sherry only account for half of their stylistic offerings...all of which are outstanding after-meal tipples, but also pair well with many desserts The sweet white wine base that is used to ramp up those sweet sherries is comprised mainly of Pedro Ximénez and

Moscatel grape wines which are extremely sweet, honeyed and pungent table wines on their own. Pedro Ximénez is doubly-dangerous because it can be, and typically is, made into a 'straw'/'passito' wine which we talked about earlier in the chapter! Air-dried to raisins, concentrating the sugars and its dark-berry skin color, PX packs a molasses punch to any oloroso base it's added to.

Sherry really is the most versatile of our fortified wines for mealtime: try the Finos and Manzanillas as aperitifs with seafood, shellfish, almonds, olives, hell really any Mediterranean tapas fare. Olorosos are on the money with cream-based soups and such. And then savor the sweet Cream sherries with desert, or as dessert itself. While you could easily spend the rest of your days fortifying your stomach and soul with the fortified wine heavyweights of Port and Sherry, they are actually not alone in their fortress of fortified solitude....

Other fun fortified friends

There are a few fun other fortified fellows that you may have heard passing reference to in wine shops, tall tales, and historical stories. Of much smaller production, much smaller production areas, and much smaller recognition to the general public, you may want to go ahead and push your palate to explore these often misunderstood wines. All have fortification with distilled neutral spirits as a common denominator, although for most of them it was initially done to preserve the wine on long ship voyages more-so than a stylistic creative endeavor...such as Madeira and Marsala.

- ❖ **Madeira** is a blended and fortified Portuguese wine made in the Madeira Islands. It used to be the wine of choice for the rich and famous of the 13 Original Colonies! Hell yeah! Jefferson, Washington, and the whole funky bunch used to love the stuff! It was heavily fortified to survive the long and extremely hot voyage below decks on the wooden ships plying the Atlantic back in the day. Apparently there was something about the heat, the rocking of the ship, and the wine itself that created a special libation in those barrels that has stood the test of time. Because it is essentially 'cooked' with heat and intentionally oxidized, the wine cannot 'go bad' after you open the bottle, and can age for centuries... or indefinitely. Supposed to be great with turtle soup. Damn! Who eats that? Oh yeah: the rich and famous of the 13 Original Colonies.
- ❖ **Marsala** is a wine produced in the region surrounding the Italian city of Marsala in Sicily. It is aged much like sherry in the solera system, and comes in both sweet and dry versions. I still mostly just use it for cooking, as sweet Marsala is the totally kick-ass base for a solid French onion soup. The hell with turtles!
- ❖ **Muscat de Beaumes de Venise** is a sweet, fortified wine from the eastern central region of the southern half of the Rhône Valley. Obviously, made by beefing up a Muscat based wine with some serious brandy. Good Rhône burn!

❖ **Vermouth** I will throw in just for kicks, as I can't imagine anyone actually drinking this stuff as an aperitif or a digestif, but it is a fortified wine from Italy. Flavored with aromatic herbs and spices such as cardamom, cinnamon, marjoram and chamomile, Vermouth is mostly used nowadays to make dry Martinis and sweet Manhattans and other old-school mixed drinks that James Bond would enjoy. Comes in sweet and dry versions, and be forewarned: it may be wine and it may be dry, but that stuff is bitter! Try it sometime to find out....or better yet dare someone else to try it and watch their reaction.

Cheers! Have some Bubbles in your Bubbly!

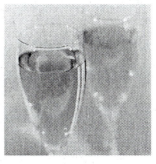

Ahhhhh....and now its time to raise a glass and make a toast as we finish up this chapter, and this entire book! And what better wine to toast with than Champagne! C'est Magnifique!!! The king of sparkling wines! The last section, of the last chapter, will deal with the last major type of wines we have to cover: the sparklers! Sparkling wines come in a wide variety of styles and sweetness, so they can make excellent aperitifs, dessert compliments, or sipping wines on their own. They pair well with virtually all lighter seafood and shellfish and mussels, most light and creamy cheeses, nuts and fruits across the board, and an array of desserts too. Sparklers are a really versatile bunch!

This is the easiest category to explain too! Sparkling wine is simply a wine that is charged with carbon dioxide....making it fizzy! Just like soda pop! Just shake the bottle up real hard and then pop the top and...whee! Wine fountain!

Okay, seriously, do not ever, ever, ever do that with sparkling wine. Unless you just won the Superbowl or a Nascar race. On second thought, don't even do it then. All that CO_2 gas charged into the wine means that the beverage is under high pressure, and the cork quickly becomes a high speed projectile which can do some serious damage to an eye. You don't want to have to go through life explaining to people that you lost your eye to a bottle of fizzy wine. That's just too uncool. But back to point....

Where does this CO_2 come from, and how does it get into the wine? Well actually, CO_2 is a natural bi-product of fermentation, so it's presence is no great mystery: as yeasts eat the sugars in grape juice, they spit out alcohol, a fair bit of heat, and carbon dioxide. The wine maker typically allows the CO_2 to escape from the vat or barrel as it is being produced. For most of wine-making history, the presence of bubbles has been a really bad thing.... a flaw! It is an indication that the fermentation was not complete before the wine was put into a bottle or a barrel. If the fermentation continues in a closed container, then the yeasts keep working, keep spitting out alcohol and heat and CO_2 until the pressure builds up enough and...KABOOM! Bottle explodes. Epic wine fail.

But with the advent of thicker and better glass, cork technology, and a greater understanding of how to more completely control the fermentation process, winemakers in the last couple hundred years have figured out how to successfully capture and contain that CO_2 in just the right amounts so as not to burst the bottle. And thus they started to intentionally craft this effervescent bubbly beverage which we have come to associate with celebrations, toasts, good times, and romantic encounters....lifestyles of the rich and famous!

The manner in which a winemaker decides to get those CO_2 bubbles trapped in the wine has a lot to do with the quality level and price of the finished wine. The most famous way to do this is the 'traditional method' or *méthode champenoise* aka the Champagne method, which of course is named for the most famous sparkling wine producing region of France....so famous, in fact, that we have come to associate the name Champagne as a generic term applying to all sparkling wine, which is NOT the case. More on that in a second. Back to the method....

Side note: the concept that Champagne/sparkling wine is a 'special' beverage for the 'special' times in the lives of 'special' people is a totally fabricated marketing ploy created by the Champagne sellers in the early 20th century in order to sell more wine. They were geniuses, and it worked....and it still does. Why else would west coast rappers need a $500 bottle of Cristal to pour out to the homies? Its all hype based on supposed exclusivity. Kind of like diamonds.

In this traditional method, the regular fermentation runs its course and produces a dry white wine. The winemaker may age the wine in barrels for a period of time to gain flavors, or he may proceed to the next step immediately. That next step entails blending and putting the wine into bottles, and simultaneously adding a bit of sugar and yeast to the mixture, which re-activates the fermentation process! This is called the secondary fermentation. Thus, the yeast will go back to work, producing some more alcohol and CO_2....but now it is trapped in the bottle! Genius! We got bubbles!

Now there are bunch of other labor-intensive steps in this process that I'm going to leave out for now, because you have the gist of how it all works. But I do want you to know a little more of the true Champagne nomenclature. For starters, true Champagne is ONLY from the Champagne region of France, is ONLY made by the *méthode champenoise*, and ONLY uses three specific grapes: Pinot Noir, Pinot Meunier and Chardonnay. This is the classic recipe. If all three of thee factors are not met, then it is just a sparkling wine. With me so far? Sweet! Oh, speaking of sweet, check the inset box for Champagne terms to know when choosing your bottle.

The winemaker decides how much sugar is to be added in that secondary fermentation stage, which ultimately affects the sweetness level of the wine. These terms which appear on every bottle of Champagne alert you the style of the wine:
- Brut nature: .0 to .5% residual sugar. Taste bone dry
- Brut: .5 to 1.5% residual sugar. Taste dry with no perceptible sweetness
- Extra Dry: 1.2 to 2% residual sugar; Tastes slightly sweet.
- Sec: 1.7 to 3.5% residual sugar. Tastes sweet to medium sweet.
- Demi-Sec: sweet, 3.3 to 5% residual sugar. Taste very sweet. (dessert)
- Doux: over 5% residual sugar. Exceptionally sweet. (dessert)

Given its reputation and the labor-intensive nature of this highly regulated method, Champagnes are typically the most expensive sparklers on the market. But, there are other, less expensive ways to get that CO_2 in the bottle: 1)Just blast it into the beverage from a tank under pressure, which is how soda pop and really cheap sparkling suprmarket swill is made, 2)the 'transfer method' in which the secondary fermentation occurs in bottles, but then the bottles are all poured into a tank and mixed together and filtered for consistency before re-bottling, and 3)the 'Charmat process' in which the wine undergoes the secondary fermentation in bulk tanks, and is bottled under pressure. These last two are how most bulk sparklers are made.

But the Champagne district of northern France does not have the sparkling market cornered my friends! Far from it! Every wine-producing country, and most wine-producing regions within each country, produce some sort of sparkling wine. Let's finish off France first: Sparkling wines labeled as **Crémant** are actually produced using the traditional method, but are from outside the Champagne region...so you might see a Crémant de Bourgogne or a Crémant de Loire. Any other sparkling wine from France that does not use the traditional Champagne method are labeled **Vins Mousseux**. On top of that, if you see a label that says **Blanc de Blanc**, that is a sparkling wine that is 100% Chardonnay, while a **Blanc de Noir** is usually 100% Pinot Noir or Pinto Meunier, or a blend of the two. You can also find lots of Rosé sparklers too.

Other countries use other grapes and other names for their sparklers, and here are a few of note:
- ❖ **Cava** is the term for Spanish sparklers produced in the méthode champenoise
- ❖ **Espumante** is Portuguese sparkler
- ❖ **Sekt** is the German and Austrian name for sparkling wine.
- ❖ **Prosecco** is a dry sparkling Italian wine made from a variety of white grape of the same name in the Veneto region.
- ❖ **Asti** or **Asti Spumante** is a sparkler made from the Moscato grape in the province of Asti, Italy.
- ❖ **Moscato d'Asti** is a slightly sweeter/lightly sparkling version of Asti. It's kind of just a little fizzy.

The New World wine countries produce a wide variety of sparklers as well, from an even wider variety of grapes. In Australia and California they are experimenting a lot with producing sparkling reds....specifically I have tried many a sparkling Shiraz—an experience I hope to never repeat. But you may love it! So give them, and all of these sparkling wines, a shot at your palate. They are all great fizzy fun!

Fin

In summary, all of the wines listed in this chapter are fantastic for before and after the meal food pairings. Or as desserts on their own. Most of the very sweet table wines like Pedro Ximénez, the Trockenbeerenauslese Rieslings, the Tokays, very sweet Sherries, and big Ports are over-the-top desserts by themselves. Just grab a glass and sit by the fire and enjoy the adult candy in a glass!

Others like the Muscats, the semi-sweet Champagnes and Sherries, or a Vin de Paille are fantastic to match with the dessert itself. As a general rule, the wine should be sweeter than the food it is served with. I have heard that a perfectly ripe peach is the ideal partner for many dessert wines, whereas it is challenging to match dessert wines with really sweet and rich chocolate-based confections. (To be honest, and solid dry red table wines go best with the chocolate spectrum.) On the other hand, most baked goods are a totally awesome accompaniment to the sweet wines, particularly with a little bitterness like the almond biscuits that are dunked in Vin Santo. Wow. And a few of these sweeties are actually over the top with really rich, savory dishes like the foie gras that is a traditional partner to Sauternes, or turtle soup with Madeira.

But don't stop there! The dry Sherries, dry Champagnes, and dry white Ports are the aperitif line-up of champions! You have not lived until you sit down in the sunshine with some almonds, olives and fresh baked hard crusty bread alongside a bottle of chilled Fino Sherry. Champagne is so fantastic with fresh fruit....one of the few wines that is. But both are also splendid with seafood! So many styles of Sherry, so many different kinds of Port, Sparklers of endless variety, and such a diverse range of dessert wines. Won't you drink some now?

It's time you classed up your act and pushed your palate to explore these 'other' wines for yourself. Many are misunderstood, most are untried, and virtually are undervalued for the well-crafted wine expressions that they are. So go! Explore! Enlighten! Entertain! Experiment! Experiment! Experiment!

Final, final note: You may have noticed that there are not a lot of specific "New World' wines listed in this chapter by name. Let me assure you that is not for lack of offerings, because there are tons of fortified and sparkling and dessert wines from these areas too. Its just that the "old World', old school wine producing regions of Europe are the ones who originally invented all these styles, and have crafted these wines for centuries under their perspective stylistic/regional names....like Vin Santo, Port, Tokay, Champagne, Sauternes....

The New World has of course mimicked and copied these wines, and perhaps put their regional spin on them a bit, but they are, in essence, local duplications of the European styles. These classic European wine styles from specific wine regions have been so wildly successful, that we have come to associate their names as a generic term representing that type of wine worldwide.....thus Champagne, Sherry and Port can now be found the world over as Australian Port-style wine, South African Sherry, and California Champagne.

Such is the exciting an ever-increasing diversity of wine in our world! Drink it now!

epilogue

These are the closing statements to a book about wine. But now that you are a wine drinker, the story will never really end…because you are just now getting started! Enjoying wine is a lifelong endeavor, and hopefully after completing this guide you are now equipped to start the long journey. And what a great trip it will be…

All I can stress to you is this: what a damn great time it is to be drinking wine!

For you newbies to the wine drinking world, you simply could not have picked a better time to get into the game. Why? For the first time in history since the fall of the Roman Empire, wine is becoming a democratic beverage again. See, back in Greek and Roman times, wine was the drink of the masses. Everyone drank wine. All peoples, of all colors, of all classes and every status were wine consumers. Wine was widely available, costs affordable, and the beverage was an ingrained component of diet and culture. Dionysus and Bacchus reigned supreme.

But since that egalitarian-drinking time, wine has become known more and more as the beverage to an exclusive class: the upper crust rich and famous. To be sure, folks actually in wine producing areas still drank wine, but availability and cost became effective barriers and thus beer (and later distilled spirits) became the drink of choice for the masses. This situation seemed to reinforce the belief that wine was only for the 'special' people like royalty, bluebloods, and snobs of all sort. What did rich people drink? Wine! What did common people drink? Ale and gin! Filthy damn peasants!

Think about it. The same situation still exists today here in 21st century America. As I alluded to in the Intro, most folks associate wine drinking with an elite class of educated, wealthy cosmopolitan types. And the masses still quaff largely flavorless light beers and hugely alcoholic cocktails. Wine to the average Joe is a beverage that costs too much money to acquire, and requires too much knowledge to understand. But now back to point…

Because that just ain't so anymore. Wine production has simply exploded across the planet. Advances in wine and grape technologies have increased productions areas, amounts of production, and stability of the product. Increases in international trade and shipping technologies now allow all of us, everywhere, to have access to all the wines of the world, everyday. Increased competition among all of the old and new wine producing areas equates to market prices falling.

Put two and two and two together here and you should come up with increased consumer choice, increased availability, and all at a decreased price! Sweet! Back to the days of democratic wine drinking. It really is time to drink this now!

On top of that, with so many new wine regions coming on-line with unique new terroirs, as well as with increased stylistic experimentation, wine style options are simply going to exponentially expand well into the future. As if there weren't enough wine options already, international wine styles and flavors and aromas are set to explode in varietal and blended expressions. One need only consider the fact that you could now easily acquire a Cabernet Sauvignon from Bordeaux, California, Australia, Chile, Sicily, South Africa, Georgia, New York, China or Bulgaria. It boggles the mind to think of the stylistic differences of just this one grape from different locations around the globe. Once you start factoring in winemaker influences, the variation increases to infinity. Now more than ever, there is a wine out there to please every palette across the planet!

So many choices. So many flavors. So many aromas. So much culture. So many positive health benefits. So many positive social benefits.   So much a component of a full, rich life…

Let me end with a line I started with: Wine is the beverage of moderation, of education, of socialization, and of civilization—and that is no exaggeration.

Won't you drink this now?

Photo Credits

Intro
Page 1
 Image archive ID: 222170
 Description: Two Glasses of Wine
 Credit: PhotoDisc/Getty Images
 Source: PhotoDisc, Inc./Getty Images
 Vendor photo ID: 48322.JPG

Page 6
 Image archive ID: 49485
 Description: Grapes and Wine
 Credit: PhotoDisc, Inc.
 Source: PhotoDisc, Inc./Getty Images
 Vendor photo ID: null
 File name: 30330.JPG

Lesson 1
Page 8
 Image archive ID: 39225
 Description: Cheese Plate
 Credit: PhotoDisc, Inc.
 Source: PhotoDisc, Inc./Getty Images
 Vendor photo ID: null
 File name: 8168.JPG

Page 10
 Image archive ID: 278934
 Description: Cheese and Crackers
 Credit: Berit Myrekrok/Digital Vision/Getty Images
 Source: Digital Vision
 Vendor photo ID: dv639051_22.jpg

Page 11
 Image archive ID: 237582
 Description: Wineglasses
 Credit: Fred Lyons/Cole Group/PhotoDisc
 Source: PhotoDisc, Inc./Getty Images
 Vendor photo ID: FD001551_7.jpg
 File name: FD001551_7.jpg

Lesson 2
Page 15 Images courtesy of Riedel Stemware

Page 16 Image courtesy of John Boyer

Page 17 Image courtesy of Riedel Stemware

Page 18 Image courtesy of Riedel Stemware

Page 22
 Image archive ID: 278934
 Description: Cheese and Crackers
 Credit: Berit Myrekrok/Digital Vision/Getty Images
 Source: Digital Vision
 Vendor photo ID: dv639051_22.jpg

Lesson 3
Page 23 Image courtesy of Riedel Stemware

Page 24 Image courtesy of Riedel Stemware

Page 26 Image courtesy of Riedel Stemware

Lesson 4
Page 31
Image archive ID: 49485
Description: Grapes and Wine
Credit: PhotoDisc, Inc.
Source: PhotoDisc, Inc./Getty Images
Vendor photo ID: null
File name: 30330.JPG

Page 35
Image archive ID: 43945
Description: Barrels of Wine
Credit: PhotoDisc, Inc.
Source: PhotoDisc, Inc./Getty Images
Vendor photo ID: null
File name: 19068.JPG

Page 38
Image archive ID: 41233
Description: Man Taking Wine from a Cask in a Wine Cellar
Credit: PhotoDisc, Inc.
Source: PhotoDisc, Inc./Getty Images
Vendor photo ID: null
File name: 12021.JPG

Page 40
Image archive ID: 284466
Description: Crackers on a Plate With Red Wine in a Wineglass
Credit: Digital Vision/Getty Images
Source: Digital Vision
Vendor photo ID: dv1354003.jpg

Lesson 5
Page 43
Image archive ID: 279025
Description: Pouring White Wine Into Glass
Credit: Berit Myrekrok/Digital Vision/Getty Images
Source: Digital Vision
Vendor photo ID: dv639075_22.jpg

Page 48
Image archive ID: 222186
Description: Glass of White Wine
Credit: PhotoDisc/Getty Images
Source: PhotoDisc, Inc./Getty Images
Vendor photo ID: 48326.JPG
File name: 48326.JPG

Lesson 6
Page 49 Image courtesy of John Boyer

Page 52
Image archive ID: 20896
Description: Spilled red wine
Credit: Corbis Digital Stock
Source: Corbis Digital Stock
Vendor photo ID: FPE0100E
File name: FPE0100E.JPG

Page 53
Image archive ID: 35153
Description: Closeup of Grapes
Credit: PhotoDisc, Inc.
Source: PhotoDisc, Inc./Getty Images
Vendor photo ID: 1336.JPG
File name: 1336.JPG

Page 54
Image archive ID: 237582
Description: Wineglasses
Credit: Fred Lyons/Cole Group/PhotoDisc
Source: PhotoDisc, Inc./Getty Images
Vendor photo ID: FD001551_7.jpg
File name: FD001551_7.jpg

Page 56
Image archive ID: 37265
Description: Close-up of Grapes
Credit: PhotoDisc, Inc.
Source: PhotoDisc, Inc./Getty Images
Vendor photo ID: null
File name: 5202.JPG

Lesson 7
Page 57
Image archive ID: 20892
Description: Glass of white wine
Credit: Corbis Digital Stock
Source: Corbis Digital Stock
Vendor photo ID: FPE0099E
File name: FPE0099E.JPG

Page 60
Image archive ID: 284466
Description: Crackers on a Plate With Red Wine in a Wineglass
Credit: Digital Vision/Getty Images
Source: Digital Vision
Vendor photo ID: dv1354003.jpg

Page 65
Image archive ID: 49433
Description: Wine and Grapes
Credit: PhotoDisc, Inc.
Source: PhotoDisc, Inc./Getty Images
Vendor photo ID: null
File name: 30317.JPG

Lesson 8
Page 68
Image archive ID: 222186
Description: Glass of White Wine
Credit: PhotoDisc/Getty Images
Source: PhotoDisc, Inc./Getty Images
Vendor photo ID: 48326.JPG
File name: 48326.JPG

Page 70
Image archive ID: 278942
Description: Glass of White Wine
Credit: Berit Myrekrok/Digital Vision/Getty Images
Source: Digital Vision
Vendor photo ID: dv639053_22.jpg

Page 76
Image archive ID: 279025
Description: Pouring White Wine Into Glass
Credit: Berit Myrekrok/Digital Vision/Getty Images
Source: Digital Vision
Vendor photo ID: dv639075_22.jpg
File name: dv639075_22.jpg

Lesson 9
Page 80 Image courtesy of John Boyer

Page 82
Image archive ID: 268738
Description: Cheeses
Credit: Corbis-Digital Stock
Source: Corbis Digital Stock
Vendor photo ID: 033.jpg
File name: 033.jpg

Page 84
Image archive ID: 238527
Description: Breads and Cheeses
Credit: Ed Carey/Cole Group/PhotoDisc
Source: PhotoDisc, Inc./Getty Images
Vendor photo ID: FD001787_7.jpg
File name: 341-FD001787_7.jpg

Lesson 10
Page 86
Image archive ID: 222170
Description: Two Glasses of Wine
Credit: PhotoDisc/Getty Images
Source: PhotoDisc, Inc./Getty Images
Vendor photo ID: 48322.JPG
File name: 48322.JPG

Page 88
Image archive ID: 153676
Description: The village of Ingersheim, viewed from Alsace's Route de Vin.
Credit: Robert Pierce
Source: Robert Pierce
Vendor photo ID: WineRoad.jpg
File name: WineRoad.jpg

Page 91
Image archive ID: 279109
Description: Stack of Various Cheese
Credit: Berit Myrekrok/Digital Vision/Getty Images
Source: Digital Vision
Vendor photo ID: dv639098_22.jpg

Lesson 11
Page 94
Image archive ID: 43925
Description: Grapes
Credit: PhotoDisc, Inc.
Source: PhotoDisc, Inc./Getty Images
Vendor photo ID: null
File name: 19063.JPG

Page 98 Image courtesy of John Boyer
Page 99 Image courtesy of John Boyer

Page 104
Image archive ID: 20684
Description: Pizza
Credit: Corbis Digital Stock
Source: Corbis Digital Stock
Vendor photo ID: FPE0047E
File name: FPE0047E.JPG

Lesson 12
Page 106
Image archive ID: 279057
Description: Glass of Red Wine
Credit: Berit Myrekrok/Digital Vision/Getty Images
Source: Digital Vision

Vendor photo ID: dv639084_22.jpg
File name: dv639084_22.jpg

Page 107
Image archive ID: 43929
Description: Pinot Grapes
Credit: PhotoDisc, Inc.
Source: PhotoDisc, Inc./Getty Images
Vendor photo ID: null
File name: 19064.JPG

Page 110
Image archive ID: 49485
Description: Grapes and Wine
Credit: PhotoDisc, Inc.
Source: PhotoDisc, Inc./Getty Images
Vendor photo ID: null
File name: 30330.JPG

Page 113
Image archive ID: 222158
Description: Glass of Red Wine
Credit: PhotoDisc/Getty Images
Source: PhotoDisc, Inc./Getty Images
Vendor photo ID: 48319.JPG

Page 113
Image archive ID: 255254
Description: Two young men and a young woman toasting with wineglasses in a restaurant
Credit: Purestock
Source: Purestock
Vendor photo ID: Purestock_1574R-03675.jpg
File name: Purestock_1574R-03675.jpg

Lesson 13
Page 118 Image courtesy of John Boyer

Page 119
Image archive ID: 39873
Description: Bottle of Wine
Credit: PhotoDisc, Inc.
Source: PhotoDisc, Inc./Getty Images
Vendor photo ID: null
File name: 8330.JPG

Page 126
Image archive ID: 41233
Description: Man Taking Wine from a Cask in a Wine Cellar
Credit: PhotoDisc, Inc.
Source: PhotoDisc, Inc./Getty Images
Vendor photo ID: null
File name: 12021.JPG

Lesson 14
Page 129
Image archive ID: 42221
Description: Man Shopping for Wine
Credit: PhotoDisc, Inc.
Source: PhotoDisc, Inc./Getty Images
Vendor photo ID: null
File name: 12277.JPG

Page 129 Image courtesy of John Boyer

Page 131
Image archive ID: 278810
Description: Glass of Red Wine

Credit: Berit Myrekrok/Digital Vision/Getty Images
Source: Digital Vision
Vendor photo ID: dv639012_22.jpg

Page 134
Image archive ID: 43945
Description: Barrels of Wine
Credit: PhotoDisc, Inc.
Source: PhotoDisc, Inc./Getty Images
Vendor photo ID: null
File name: 19068.JPG

Page 135 Image courtesy of John Boyer

Lesson 15
Page 139
Image archive ID: 279049
Description: Glasses of Red Wine
Credit: Berit Myrekrok/Digital Vision/Getty Images
Source: Digital Vision
Vendor photo ID: dv639081_22.jpg
File name: dv639081_22.jpg

Page 144
Image archive ID: 278934
Description: Cheese and Crackers
Credit: Berit Myrekrok/Digital Vision/Getty Images
Source: Digital Vision
Vendor photo ID: dv639051_22.jpg
File name: dv639051_22.jpg

Page 146
Image archive ID: 42221
Description: Man Shopping for Wine
Credit: PhotoDisc, Inc.
Source: PhotoDisc, Inc./Getty Images
Vendor photo ID: null
File name: 12277.JPG

Lesson 16
Page 149
Image archive ID: 278774
Description: Pouring Wine Into Glass
Credit: Berit Myrekrok/Digital Vision/Getty Images
Source: Digital Vision
Vendor photo ID: dv639002_22.jpg
File name: dv639002_22.jpg

Page 158
Image archive ID: 20856
Description: Corkscrew, cork, and glass of red wine
Credit: Corbis Digital Stock
Source: Corbis Digital Stock
Vendor photo ID: FPE0090E
File name: FPE0090E.JPG

Lesson 17
Page 160
Image archive ID: 43933
Description: Pinot and Riesling Grapes
Credit: PhotoDisc, Inc.
Source: PhotoDisc, Inc./Getty Images
Vendor photo ID: null
File name: 19065.JPG

Page 162
Image archive ID: 222158

Description: Glass of Red Wine
Credit: PhotoDisc/Getty Images
Source: PhotoDisc, Inc./Getty Images
Vendor photo ID: 48319.JPG

Page 165
Image archive ID: 43945
Description: Barrels of Wine
Credit: PhotoDisc, Inc.
Source: PhotoDisc, Inc./Getty Images
Vendor photo ID: null
File name: 19068.JPG

Page 168
Image archive ID: 42341
Description: Wine and Cheese
Credit: PhotoDisc, Inc.
Source: PhotoDisc, Inc./Getty Images
Vendor photo ID: null
File name: 12310.JPG

Lesson 18
Page 169
Image archive ID: 42293
Description: Corkscrew Removing a Cork from a Wine Bottle
Credit: PhotoDisc, Inc.
Source: PhotoDisc, Inc./Getty Images
Vendor photo ID: null
File name: 12297.JPG

Page 171
Image archive ID: 48409
Description: Oysters and White Wine
Credit: PhotoDisc, Inc.
Source: PhotoDisc, Inc./Getty Images
Vendor photo ID: null
File name: 30059.JPG

Page 174
Image archive ID: 220618
Description: Port Aspic with Duck Liver Moussiline
Credit: PhotoDisc/Getty Images
Source: PhotoDisc, Inc./Getty Images
Vendor photo ID: 48228.JPG
File name: 48228.JPG

Chapter 19
Page 177
Image archive ID: 181244
Description: Bottles of Wine and Liquor
Credit: PhotoDisc/Getty Images
Source: PhotoDisc, Inc./Getty Images
Vendor photo ID: FD000682.jpg
File name: FD000682.jp

Page 179
Image archive ID: 19468
Description: Wine cellar, Monte Carlo, Monaco
Credit: Corbis Digital Stock
Source: Corbis Digital Stock
Vendor photo ID: DEU0043E
File name: DEU0043E.JPG

Page 182
Image archive ID: 49445
Description: Banana Dessert
Credit: PhotoDisc, Inc.

Source: PhotoDisc, Inc./Getty Images
Vendor photo ID: null
File name: 30320.JPG

Page 184
Image archive ID: 20888
Description: Snifter of brandy
Credit: Corbis Digital Stock
Source: Corbis Digital Stock
Vendor photo ID: FPE0098E
File name: FPE0098E.JPG

Page 187
Image archive ID: 49469
Description: Chocolate Cake
Credit: PhotoDisc, Inc.
Source: PhotoDisc, Inc./Getty Images
Vendor photo ID: null
File name: 30326.JPG

Page 188
Image archive ID: 49437
Description: Apple Strudel
Credit: PhotoDisc, Inc.
Source: PhotoDisc, Inc./Getty Images
Vendor photo ID: null
File name: 30318.JPGEpilogue

Page 191
Image archive ID: 279017
Description: Two Glasses of Champagne
Credit: Berit Myrekrok/Digital Vision/Getty Images
Source: Digital Vision
Vendor photo ID: dv639073_22.jpg
File name: dv639073_22.jpg

Page 192
Description: Champagne
Credit: PhotoDisc, Inc.
Source: PhotoDisc, Inc./Getty Images
Vendor photo ID: null
File name: 30328.JPG

Page 194
Image archive ID: 33205
Description: Chilled Champagne
Credit: PhotoDisc, Inc.
Source: PhotoDisc, Inc./Getty Images
Vendor photo ID: OS09075.JPG
File name: OS09075.JPG

Page 196
Image archive ID: 49469
Description: Chocolate Cake
Credit: PhotoDisc, Inc.
Source: PhotoDisc, Inc./Getty Images
Vendor photo ID: null
File name: 30326.JPG

Page 199
Image archive ID: 43929
Description: Pinot Grapes
Credit: PhotoDisc, Inc.
Source: PhotoDisc, Inc./Getty Images
Vendor photo ID: null
File name: 19064.JPG